BestMasters

Mit „**BestMasters**" zeichnet Springer die besten Masterarbeiten aus, die an renommierten Hochschulen in Deutschland, Österreich und der Schweiz entstanden sind. Die mit Höchstnote ausgezeichneten Arbeiten wurden durch Gutachter zur Veröffentlichung empfohlen und behandeln aktuelle Themen aus unterschiedlichen Fachgebieten der Naturwissenschaften, Psychologie, Technik und Wirtschaftswissenschaften. Die Reihe wendet sich an Praktiker und Wissenschaftler gleichermaßen und soll insbesondere auch Nachwuchswissenschaftlern Orientierung geben.

Springer awards **"BestMasters"** to the best master's theses which have been completed at renowned Universities in Germany, Austria, and Switzerland. The studies received highest marks and were recommended for publication by supervisors. They address current issues from various fields of research in natural sciences, psychology, technology, and economics. The series addresses practitioners as well as scientists and, in particular, offers guidance for early stage researchers.

More information about this series at http://www.springer.com/series/13198

Stefan Rocktäschel

A Branch-and-Bound Algorithm for Multiobjective Mixed-integer Convex Optimization

 Springer Spektrum

Stefan Rocktäschel
Ilmenau, Germany

ISSN 2625-3577   ISSN 2625-3615   (electronic)
BestMasters
ISBN 978-3-658-29148-8   ISBN 978-3-658-29149-5   (eBook)
https://doi.org/10.1007/978-3-658-29149-5

This Springer Spektrum imprint is published by the registered company Springer Fachmedien
Wiesbaden GmbH part of Springer Nature.
The registered company address is: Abraham-Lincoln-Str. 46, 65189 Wiesbaden, Germany

This book is dedicated to my family and friends.
Thank you for being there!

Contents

1 Introduction

Mixed-integer optimization problems (MIP) appear in a variety of applications like in economics or engineering. One example is the uncapacitated facility location problem studied by Günlük, Lee, Weismantel [9], where integer variables are used to model the decision for a facility, whether it should be built or not. Additionally, there are continuous variables which state the percentage of the respective customers' demands which is met by any given facility. The objective hereby is to decide which facilities to build in order to minimize costs. Mixed-integer optimization problems have been studied for example in [13] and [2]. There are already some solvers for these optimization problems [10], [1] ..

Another class of optimization problems that are of interest in many applications are multiobjective optimization problems (MOP). Hereby, multiple objective functions have to be minimized simultaneously. In general, there is no point, i.e. choice of variables that minimizes all objective functions at the same time. As a result, there is another concept of optimality used for this class of optimization problems than we know from scalar optimization. Multiobjective optimization problems have been studied in [11] and [6] for example and there are already solvers for these problems [7], [16].

In this book, we use techniques from both of the above classes of optimization problems and study multiobjective mixed-integer convex

© Springer Fachmedien Wiesbaden GmbH, part of Springer Nature 2020
S. Rocktäschel, *A Branch-and-Bound Algorithm for Multiobjective Mixed-integer Convex Optimization*, BestMasters, https://doi.org/10.1007/978-3-658-29149-5_1

optimization problems (MOMICP). This is a very important class of optimization problems since it generalizes the concepts of both of the above classes. Considering (MOMICP), we are able to add additional objective functions, which we want to minimize, for example to the uncapacitated facility location problem. One could derive an additional objective function for this case by introducing a parameter for each facility that measures the negative impact on the environment that occurs, if the respective facility is built. Our additional objective could be to minimize the total negative impact on the environment with our building plan for the facilities.

For solving (MOMICP) we are interested in finding the set of *efficient points*. These are points for which there exists no other feasible point that is better or equally good in all objectives. We are also interested in finding *nondominated points*, which are the images of the efficient points. This is one main difficulty of (MOMICP) in comparison to (MIP). For (MIP) it is possible, at least theoretically, to completely enumerate all combinations of integer variables and solve a continuous optimization problem for them. Afterwards, one can compare all obtained minima and the smallest solves the (MIP). Despite this is not a 'good' approach for solving (MIP), it is even worse for (MO-MICP). The reason is that we would have to enumerate and solve multiobjective optimization problems that are much harder to solve, because we get a whole set of nondominated points for each of these optimization problems instead of a unique minimal function value and then we would have to compare all these nondominated sets. So we note that this naive approach seems not practicable at all. There are algorithms for solving (MOMICP) as can be seen in [5]. However, the approach introduced in that paper is only heuristic. Thus, our aim is to develop an algorithm that obtains efficient and nondominated points of the optimization problem (MOMICP). Hereby, we use concepts and techniques that are known from mixed-integer optimization and from global multiobjective optimization.

2 Theoretical Basics

In this chapter, we introduce the basic concepts of multiobjective optimization. We introduce basic definitions and derive a concept of optimality for multiobjective optimization problems. Based on this, we formulate the central optimization problem that we study throughout this book and introduce a relaxed optimization problem that we use in order to solve the central optimization problem. Throughout this book $\mathbb{N}$ denotes the set of natural numbers, $\mathbb{Z}$ the set of integers, $\mathbb{R}$ the set of real numbers and $\mathbb{R}_+$ the set of nonnegative real numbers.

2.1 Basics of multiobjective optimization

In this section, we set the basis for comparing vectors in $\mathbb{R}^p$. Therefore, we use the pointed convex cone $K = \mathbb{R}^p_+$, which induces a partial order relation $\leqslant$ on $\mathbb{R}^p$ defined by

$$z^1 \leqslant z^2 :\Leftrightarrow z^2 - z^1 \in \mathbb{R}^p_+$$

for elements $z^1, z^2 \in \mathbb{R}^p$, also called the componentwise ordering in $\mathbb{R}^p$. Obviously, it holds

$$z^1 \leqslant z^2 \Leftrightarrow z^1_i \leqslant z^2_i \text{ for all } i \in \{1, ..., p\}.$$

© Springer Fachmedien Wiesbaden GmbH, part of Springer Nature 2020
S. Rocktäschel, *A Branch-and-Bound Algorithm for Multiobjective Mixed-integer Convex Optimization*, BestMasters, https://doi.org/10.1007/978-3-658-29149-5_2

Note that the partial order $\leqslant$ on $\mathbb{R}^p$ is not a total order relation for $p \geqslant 2$. Additionally, we use the following notations:

- $z^1 \lneqq z^2 :\Leftrightarrow z^2 - z^1 \in \mathbb{R}^p_+ \backslash \{0_p\}$ and

- $z^1 < z^2 :\Leftrightarrow z^2 - z^1 \in \text{int}(\mathbb{R}^p_+)$,

where $\text{int}(\mathbb{R}^p_+)$ is *the interior* of the set $\mathbb{R}^p_+$.
The following definition allows us to consider projections of subsets $Z \subseteq \mathbb{R}^p$.

Definition 2.1. *Let the set $Z \subseteq \mathbb{R}^p$ be nonempty and $j \in \{1, ..., p\}$. We define the j-projection of Z as*

$$Z_j := \{z_j \in \mathbb{R} \mid \exists \tilde{z} \in Z : \tilde{z}_j = z_j\}.$$

Remark 2.2. *We recall that for nonempty and compact set $Z \subseteq \mathbb{R}^p$ all projections $Z_1, ..., Z_p$ of Z are nonempty and compact.*

The following definition generalizes the concepts of minimality and maximality from scalar optimization to multiobjective optimization and introduces the ideal- and the anti-ideal point of a set $Z \subseteq \mathbb{R}^p$.

Definition 2.3. *Let $Z \subseteq \mathbb{R}^p$ be a given set. An element $z^* \in Z$ is called*

- *a minimal element of Z, if $Z \cap (\{z^*\} - \mathbb{R}^p_+) = \{z^*\}$ and*

- *a maximal element of Z, if $Z \cap (\{z^*\} + \mathbb{R}^p_+) = \{z^*\}$.*

Additionally,

- *the set of minimal elements of Z is given by*

$$\min(Z) := \{z^* \in Z \mid Z \cap (\{z^*\} - \mathbb{R}^p_+) = \{z^*\}\}$$

and

- *the set of maximal elements of Z is given by*

$$\max(Z) := \{z^* \in Z \mid Z \cap (\{z^*\} + \mathbb{R}_+^p) = \{z^*\}\}.$$

Furthermore, if Z is nonempty and compact, we define

- *the ideal point $\underline{\min}(Z)$ of Z as the element $z^* \in \mathbb{R}^p$ with*
  $z_j^* = \min(Z_j) = \min\{z_j \in \mathbb{R} \mid z \in Z\}$ *for all $j \in \{1, ..., p\}$ and*

- *the anti-ideal point $\overline{\max}(Z)$ of Z as the element $z^* \in \mathbb{R}^p$ with*
  $z_j^* = \max(Z_j) = \max\{z_j \in \mathbb{R} \mid z \in Z\}$ *for all $j \in \{1, ..., p\}$.*

Thereby, $\min(Z_j)$ and $\max(Z_j)$ for $j \in \{1, ..., p\}$ refer to the scalar minimum and maximum of the set Z_j.

Remark 2.4. *The above definitions of $\underline{\min}(Z)$ and $\overline{\max}(Z)$ are well defined, because the respective minima and maxima exist due to the compactness and non emptiness of the set Z and hence, of the projections.*
Furthermore, it is easy to see that $\{z \in Z \mid z \leqslant z^*\} = \{z^*\}$ holds for all $z^* \in \min(Z)$. Analogously, $\{z \in Z \mid z^* \leqslant z\} = \{z^*\}$ holds for all $z^* \in \max(Z)$.
Additionally,

$$\begin{aligned}
\min(-Z) &= \{z \in \mathbb{R}^p \mid (-Z) \cap (\{z\} - \mathbb{R}_+^p) = \{z\}\} \\
&= \{-z \in \mathbb{R}^p \mid (-Z) \cap (\{-z\} - \mathbb{R}_+^p) = \{-z\}\} \\
&= \{-z \in \mathbb{R}^p \mid Z \cap (\{z\} + \mathbb{R}_+^p) = \{z\}\} \\
&= -\{z \in \mathbb{R}^p \mid Z \cap (\{z\} + \mathbb{R}_+^p) = \{z\}\} \\
&= -\max(Z).
\end{aligned}$$

This equation can be used to reformulate maximizing problems as minimizing problems and vice versa.

The following example illustrates the above definitions.

Example 2.5. *We consider the set $Z = \{z \in \mathbb{R}^2 \mid ||z||_2 \leqslant 1\}$.*
The minimum of Z is given by $\min(Z) = \{z \in \mathbb{R}^2 \mid ||z||_2 = 1, z_1 \leqslant 0, z_2 \leqslant 0\}$ and the maximum of Z is given by $\max(Z) = \{z \in \mathbb{R}^2 \mid ||z||_2 = 1, z_1 \geqslant 0, z_2 \geqslant 0\}$. Furthermore, it holds $\underline{\min}(Z) = (-1, -1)^\top$ and $\overline{\max}(Z) = (1, 1)^\top$. The sets and points are illustrated in Figure 2.1.

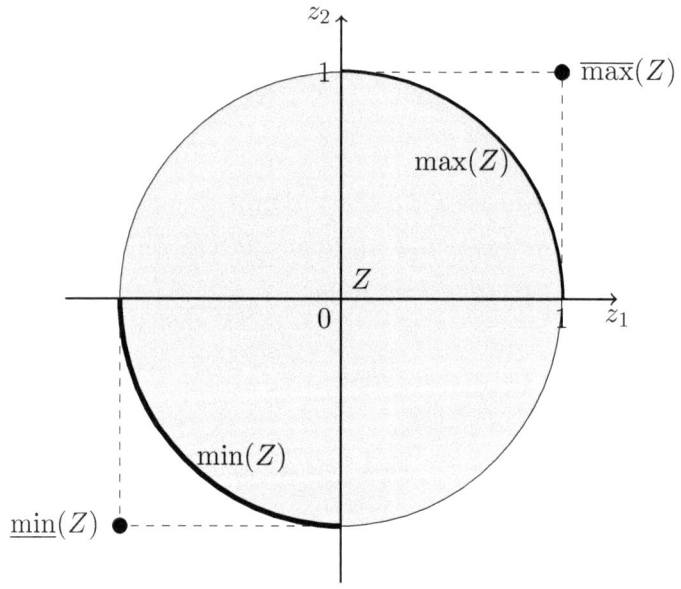

Figure 2.1: the sets and points from Example 2.5

2.2 The central multiobjective mixed-integer optimization problem

In this section, we introduce basic notations and definitions that allow us to formulate the central optimization problem of this book. At first, we recall the definition of intervals and generalize this concept for the multidimensional case.

At first, we introduce intervals and boxes [15].

Definition 2.6.

(i) For $\underline{b}, \bar{b} \in \mathbb{R}$ with $\underline{b} \leqslant \bar{b}$ we define the nonempty, compact interval B by

$$B = [\underline{b}, \bar{b}] := \{b \in \mathbb{R} \mid \underline{b} \leqslant b \leqslant \bar{b}\}.$$

(ii) We denote the set of all of these intervals by $\mathbb{IR} := \{[\underline{b}, \bar{b}] \subseteq \mathbb{R} \mid \underline{b}, \bar{b} \in \mathbb{R}, \underline{b} \leqslant \bar{b}\}$.

Now, we introduce the definition of boxes that are a generalization of intervals for the multidimensional case.

Definition 2.7.

(i) For $\underline{b}, \bar{b} \in \mathbb{R}^r$ with $\underline{b} \leqslant \bar{b}$ we define the nonempty, compact, r-dimensional Box B by

$$B = [\underline{b}, \bar{b}] := \{b \in \mathbb{R}^r \mid \underline{b} \leqslant b \leqslant \bar{b}\} = \bigtimes_{i=1}^{r} [\underline{b}_i, \bar{b}_i].$$

(ii) We denote the set of all of these r-dimensional boxes in $\mathbb{R}^r$ by $\mathbb{IR}^r$.

(iii) Let $B, \tilde{B} \in \mathbb{R}^r$ be boxes with $\tilde{B} \subseteq B$. Then we call $\tilde{B}$ a subbox of B.

We introduce notations for certain parameters of boxes.

Definition 2.8. *For $B = [\underline{b}, \overline{b}] \in \mathbb{R}^r$ we define*

(i) the infimum of B as $\inf(B) := \underline{b}$,

(ii) the supremum of B as $\sup(B) := \overline{b}$,

(iii) the midpoint of B as $\mathrm{mid}(B) := \frac{\underline{b}+\overline{b}}{2}$ and

(iv) the width of B as $\mathrm{wid}(B) := \max\limits_{i \in \{1,\ldots,r\}} (\overline{b}_i - \underline{b}_i)$.

Furthermore, we call elements $b \in B$ with $b_i \in \{\underline{b}_i, \overline{b}_i\}$ for all $i \in \{1, \ldots, r\}$ vertices of B.

Obviously, boxes are convex sets.

Moreover, we will use the following notations. Let $B \subseteq \mathbb{R}^r$ be a set and $f \colon B \to \mathbb{R}$ a function that is twice continuously differentiable on an open superset $\overline{B}$ of B. We define $f(\tilde{B}) := \{f(b) \mid b \in \tilde{B}\}$ for $\tilde{B} \subseteq B$. Furthermore, for an element $b \in \overline{B}$ we denote the gradient of f in b as $\nabla f(b)$ and the Hessian of f in b as $H_f(b)$. In addition, $\lambda_{\min_f}(b)$ denotes the smallest eigenvalue of $H_f(b)$. Since f is twice continuously differentiable $\lambda_{\min_f}(b)$ is well defined, because $H_f(b)$ is symmetric for all $b \in \overline{B}$ and hence, every eigenvalue of $H_f(b)$ is real.

The following assumptions will be used to define the central optimization problem of this book and to prove associated theoretical results.

Assumption 2.9. *Let $X = [\underline{x}, \overline{x}] \in \mathbb{R}^m$, $Y = [\underline{y}, \overline{y}] \in \mathbb{R}^n$ be boxes for $m, n \in \mathbb{N}$ and $B := X \times Y \in \mathbb{R}^r$ with $r := m + n$. Additionally, we assume $\underline{y}, \overline{y} \in \mathbb{Z}^n$.*

Furthermore, let the functions $f \colon B \to \mathbb{R}^p$ and $g \colon B \to \mathbb{R}^q$ for $p, q \in \mathbb{N}$, $p \geqslant 2$ be convex and twice continuously differentiable on an open superset $\overline{B}$ of B.

Remark 2.10. *For our purposes, the assumption $\underline{y}, \overline{y} \in \mathbb{Z}^n$ can be made without loss of generality, since we could simply round these vectors up or down respectively and would obtain equivalent optimization problems.*
We will also make an additional assumption on regularity of certain optimization problems in Assumption 3.9 later on.

Using the notation of Assumption 2.9, we introduce notations for subsets of B that will be used later on.

Definition 2.11. *Let Assumption 2.9 be fulfilled and let $\tilde{B} = \tilde{X} \times \tilde{Y} \subseteq B$ with $\tilde{X} \in \mathbb{R}^m$ and $\tilde{Y} \in \mathbb{R}^n$ be a given subbox of B. We define*

$$\tilde{B}^g := \{(x, y) \in \tilde{X} \times \tilde{Y} \mid g(x, y) \leqslant 0_q\} \ and$$

$$\tilde{B}^{\mathbb{Z}} := \{(x, y) \in \tilde{X} \times \tilde{Y} \mid y \in \mathbb{Z}^n\}, \ as \ well \ as$$

$$\tilde{B}^{g,\mathbb{Z}} := \tilde{B}^g \cap \tilde{B}^{\mathbb{Z}}.$$

Now, we are able to formulate the central optimization problem that we are going to study in this book. Under Assumption 2.9 we are going to study the multiobjective mixed-integer convex problem

$$\begin{aligned} &\min \ f(x, y) \\ &\text{s.t. } \ g(x, y) \leqslant 0_q \\ &\qquad (x, y) \in X \times Y = B \\ &\qquad y \in \mathbb{Z}^n. \end{aligned} \qquad \text{(MOMICP)}$$

Using the notations from above, we can also write (MOMICP) as

$$\min f(b)$$
$$\text{s.t. } b \in B^{g,\mathbb{Z}}.$$

Remark 2.12. *Although (MOMICP) is a generalization of multi-objective convex optimization problems, this approach cannot be used in order to solve the interesting class of optimization problems defined by*

$$\min y^\top \tilde{f}(x)$$
$$\text{s.t. } y^\top \tilde{g}(x) \leqslant 0_q \qquad\qquad (\mathrm{P}_y)$$
$$x \in X$$
$$y \in \{0,1\}^n$$

with $X \in \mathbb{R}^m$, $\tilde{f} \colon X \to \mathbb{R}^{n \times p}$ and $\tilde{g} \colon X \to \mathbb{R}^{n \times q}$, where $\tilde{f}_{i,j}$ and $\tilde{g}_{i,k}$ are convex and twice continuously differentiable on an open superset $\overline{X}$ of X for all $i \in \{1, ..., n\}$, $j \in \{1, ..., p\}$, $k \in \{1, ..., q\}$.
This is due to the fact that we demand convexity of f and g in Assumption 2.9. With $Y := [0,1]^n$, $B := X \times Y$, $f \colon B \to \mathbb{R}^p$ defined by $f(x,y) := y^\top \tilde{f}(x)$ and $g \colon B \to \mathbb{R}^q$ defined by $g(x,y) := y^\top \tilde{g}(x)$ we can rewrite (P_y) in the form of (MOMICP). However, the functions f and g are not necessarily convex in this case, even not if $\tilde{f}$ and $\tilde{g}$ are linear. We illustrate this in Example 2.13.

Example 2.13. *Consider $X = [-1,1]$, $n = 2$, $q = 1$ and $\tilde{g} \colon X \to \mathbb{R}^2$ with $\tilde{g}(x) := (x, -x)^\top$ in (P_y). Note that $\tilde{g}$ is linear and hence convex. Then we define $g \colon X \times [0,1]^2 \to \mathbb{R}$ by $g(x,y) := y^\top \tilde{g}(x)$.*
For $(x,y) = (-1, (0,0)^\top) \in X \times [0,1]^2$, $(x', y') = (1, (0,1)^\top) \in X \times$

$[0,1]^2$ *and* $\lambda = \frac{1}{2} \in [0,1]$ *we obtain*

$$g(\lambda(x,y) + (1-\lambda)(x',y')) = g((-\frac{1}{2}, (0,0)^\top) + (\frac{1}{2}, (0, \frac{1}{2})^\top))$$
$$= g(0, (0, \frac{1}{2})^\top)$$
$$= (0, \frac{1}{2}) \cdot \tilde{g}(0)$$
$$= (0, \frac{1}{2}) \cdot (0,0)^\top = 0,$$

but

$$\lambda g(x,y) + (1-\lambda)g(x',y') = \lambda(0,0) \cdot \tilde{g}(-1) + (1-\lambda)(0,1) \cdot \tilde{g}(1)$$
$$= \frac{1}{2}(0,0) \cdot (-1,1)^\top + \frac{1}{2}(0,1) \cdot (1,-1)^\top$$
$$= 0 - \frac{1}{2} = -\frac{1}{2} < 0.$$

Therefore, g is not convex.

Note that we could also find an example where the objective functions f of (P_y) are nonconvex in an analog way.

At least, there are possibilities that allow us to neglect the assumption of convexity of f for (MOMICP). We will outline one approach for this in Section 6.

Now, in order to derive a concept of optimality for the central multiobjective optimization problem (MOMICP), we introduce the concept efficiency for multiobjective optimization problems

$$\min \ f(b)$$
$$\text{s.t. } b \in B, \tag{MOP}$$

where $B \subseteq \mathbb{R}^r$ is a given set with $r \in \mathbb{N}$ and $f \colon B \to \mathbb{R}^p$ is a given function with $p \in \mathbb{N}$. In the following definitions we introduce the concepts of efficiency and nondominated points [11], [8].

Definition 2.14.

(i) *A point $b^* \in B$ is called efficient for (MOP), if there is no $b \in B$ with*

$$f(b) \leqslant f(b^*).$$

(ii) *Let $b, b^* \in B$ with*

$$f(b) \leqslant f(b^*).$$

Then we say b dominates b^*.

(iii) *The set of efficient points for (MOP) is called the efficient set of (MOP).*

(iv) *A set $\mathcal{L} \subseteq \mathcal{P}(\mathbb{R}^r)$ is called a cover of the efficient set E of (MOP), if $E \subseteq \bigcup \mathcal{L} := \bigcup_{\tilde{B} \in \mathcal{L}} \tilde{B}$ holds.*

Remark 2.15. *Referring to Definition 2.3, we observe that a feasible point $b^* \in B$ is efficient for (MOP) if, and only if, $f(b^*)$ is a minimal element of $f(B)$ or in other words $f(b^*) \in \min(f(B))$.*

The above definitions consider points in the pre-image space of f. Additionally, we introduce definitions for points in the image space of f.

Definition 2.16.

(i) *A point $z^* = f(b^*)$ is called nondominated for (MOP), if $b^* \in B$ is efficient for (MOP).*

(ii) *The set of all nondominated points of (MOP) is called the nondominated set of (MOP).*

(iii) For elements $z^1, z^2 \in \mathbb{R}^p$ with $z^1 \lneq z^2$ we say z^1 dominates z^2.

(iv) A set $Z \subseteq \mathbb{R}^p$ is called stable, if there are no $z^1, z^2 \in Z$ with $z^1 \lneq z^2$.

Remark 2.17. *Referring to Definition 2.3, we observe that a point $z^* \in \mathbb{R}^p$ is nondominated if, and only if, $z^* \in \min(f(B))$. Hence, the nondominated set of (MOP) is equal to $\min(f(B))$.*

2.3 A relaxation of (MOMICP)

In this section, deriving from the optimization problem (MOMICP), we introduce a relaxed optimization problem $(\mathrm{ROP}(\tilde{B}))$ for a given subbox $\tilde{B}$ of B. We use $(\mathrm{ROP}(\tilde{B}))$ to determine lower bounds for f on $\tilde{B}^{g,\mathbb{Z}}$. These lower bounds will be sets $L \subseteq \mathbb{R}^p$ of a 'simple' structure that fulfill $f(\tilde{B}^{g,\mathbb{Z}}) \subseteq L + \mathbb{R}^p_+$.

Under Assumption 2.9, the relaxed optimization problem we are interested in is given by

$$\begin{aligned} \min \; & f(b) \\ \text{s.t.} \; & b \in \tilde{B}^g. \end{aligned} \qquad (\mathrm{ROP}(\tilde{B}))$$

Obviously, the set of feasible points for (MOMICP) $B^{g,\mathbb{Z}}$ is a subset of the set of feasible points for $(\mathrm{ROP}(B))$, which is B^g. Because of this, every feasible point for (MOMICP) is feasible for $(\mathrm{ROP}(B))$ and we call the optimization problem $(\mathrm{ROP}(B))$ a *relaxation* of (MOMICP). Since there are no integer constraints tied to the relaxation $(\mathrm{ROP}(B))$, it is a multi-objective (continuous) convex optimization problem. This type of optimization problems has already been studied extensively.

3 A basic Branch-and-Bound algorithm for (MOMICP)

In this chapter, we introduce a basic algorithm for computing a 'good' cover of the efficient set of (MOMICP). The algorithm illustrates the basic procedure that we use. The idea of this Branch-and-Bound algorithm is to iteratively split the initial box B into smaller subboxes and derive lower and upper bounds for respective subproblems. Using these bounds, we can check certain criteria that allow us to discard a subbox, if it cannot contain any efficient point for (MOMICP). We also check for a termination criteria assuring that the boxes, which fulfill it, are of interest for (MOMICP) (e.g. contain efficient points). If neither of those criteria are fulfilled, the respective box will be further investigated and split again later on. After checking any box regarding these criteria, the next box to check is selected via a selection rule. The selection rule, the bisection step, the discarding test and the termination criteria, as well as the determination of lower and upper bounds for the respective subproblems and their comparison are used as black boxes for now. In the forthcoming chapters, we investigate these black boxes and introduce possible options to replace them. We will use the following notations.

- The list $\mathcal{L}_{\mathcal{W}}$ denotes the working list and contains boxes that still have to be examined.

© Springer Fachmedien Wiesbaden GmbH, part of Springer Nature 2020
S. Rocktäschel, *A Branch-and-Bound Algorithm for Multiobjective Mixed-integer Convex Optimization*, BestMasters, https://doi.org/10.1007/978-3-658-29149-5_3

- The list $\mathcal{L_S}^1$ denotes the solution list and contains boxes that fulfill the termination criteria.

- The list $\mathcal{L_{NS}}^1$ denotes the non-solution list and contains boxes that can be discarded according to the discarding test or a necessary feasibility condition.

- The list $\mathcal{L_{PNS}}$ denotes the set of potentially nondominated points.

- The list $\mathcal{L_{LUB}}$ denotes the set of local upper bounds w.r.t. $\mathcal{L_{PNS}}$.

How we manage $\mathcal{L_{PNS}}$ and $\mathcal{L_{LUB}}$ is not explicitly stated in Algorithm 1. We will go more into detail on how to manage these lists in Sections 3.4 and 3.5.

Algorithm 1 Basic Branch-and-Bound algorithm for (MOMICP)

INPUT: (MOMICP)

OUTPUT: $\mathcal{L}_{\mathcal{S}}^{1}$, $\mathcal{L}_{\mathcal{NS}}^{1}$, $\mathcal{L}_{\mathcal{PNS}}$, $\mathcal{L}_{\mathcal{LUB}}$

1: $\mathcal{L}_{\mathcal{S}}^{1} \leftarrow \varnothing$, $\mathcal{L}_{\mathcal{NS}}^{1} \leftarrow \varnothing$, $\mathcal{L}_{\mathcal{W}} \leftarrow \{B\}$

2: **while** $\mathcal{L}_{\mathcal{W}} \neq \varnothing$ **do**

3:     Select a box $\tilde{B}$ of $\mathcal{L}_{\mathcal{W}}$ via the selection rule.

4:     Delete $\tilde{B}$ from $\mathcal{L}_{\mathcal{W}}$.

5:     Bisect $\tilde{B}$ into subboxes $\tilde{B}^{1}$ and $\tilde{B}^{2}$ by applying the bisection step.

6:     **for** $j = 1, 2$ **do**

7:         **if** $\tilde{B}^{j}$ does not fulfill the necessary feasibility condition **then**

8:             Discard $\tilde{B}^{j}$ by adding it to $\mathcal{L}_{\mathcal{NS}}^{1}$.

9:         **else**

10:             Determine lower bounds of f on $(\tilde{B}^{j})^{g,\mathbb{Z}}$ and upper bounds of f on $B^{g,\mathbb{Z}}$.

11:         **if** $\tilde{B}^{j}$ can be discarded via the discarding test **then**

12:             Discard $\tilde{B}^{j}$ by adding it to $\mathcal{L}_{\mathcal{NS}}^{1}$.

13:         **else**

14:             **if** $\tilde{B}^{j}$ fulfills the termination criteria **then**

15:                 Add $\tilde{B}^{j}$ to $\mathcal{L}_{\mathcal{S}}^{1}$.

16:             **else**

17:                 Add $\tilde{B}^{j}$ to $\mathcal{L}_{\mathcal{W}}$.

18:             **end if**

19:         **end if**

20:         **end if**

21:     **end for**

22: **end while**

3.1 The selection rule

Referring to Algorithm 1, the selection rule determines, which box of $\mathcal{L}_{\mathcal{W}}$ is going to be examined in the current iteration. Since we try to find minimal elements of $f(B^{g,\mathbb{Z}})$, it would make sense to choose a subbox $\tilde{B}$ of B that provides 'small' function values in $f(\tilde{B}^{g,\mathbb{Z}})$. It is not manageable to compare these image sets among all boxes on $\mathcal{L}_\mathcal{W}$ pairwise in order to achieve this. Instead the ideal point $\underline{\min}(f(\tilde{B}^g))$ for boxes $\tilde{B} \in \mathcal{L}_\mathcal{W}$ can easily be determined and is a relatively good indicator for the function values of f on $\tilde{B}^g$ and hence, can be used as indicator for the function values of f on $\tilde{B}^{g,\mathbb{Z}}$.

Using the definition of the ideal point (Definition 2.3), it is obvious that we can obtain $\underline{\min}(f(\tilde{B}^g))$ for a box $\tilde{B} \in \mathcal{L}_\mathcal{W}$ by solving the scalar convex optimization problems

$$
\begin{aligned}
&\min \ f_i(b) \\
&\text{s.t. } b \in \tilde{B}^g
\end{aligned}
\qquad (\text{ROP}_i(\tilde{B}))
$$

with minimum value φ_i for all $i \in \{1, ..., p\}$. The ideal point $\underline{\min}(f(\tilde{B}^g))$ is then given by

$$
\underline{\min}(f(\tilde{B}^g)) = (\varphi_1, ..., \varphi_p)^\top.
$$

Our selection rule is to select a box $\tilde{B} \in \mathcal{L}_\mathcal{W}$ with the lexicographic smallest ideal point $\underline{\min}(f(\tilde{B}^g))$. If there is more than one box with the lexicographic smallest ideal point, we select the box, which we obtained by bisection (see Section 3.2) at first. Although, there are other tiebreaker rules one could apply here.

However, this selection rule is only a heuristic and does not guarantee to select a 'good' box. On the other hand, it is fairly easy to determine these ideal points and since the lexicographical order on $\mathbb{R}^p$ is total, there is exactly one ideal point that can be chosen.

Other approaches that could make sense would be to choose any ideal point that is not dominated by others or calculate a weighted sum of the components of the respective ideal points and choose one that minimizes this sum. However, these approaches are only heuristics as well and might not have a unique solution, what is the case for our approach. As a consequence, these ideas would need additional tiebreaker rules.

We can also observe that the lexicographical smallest ideal point is always nondominated by others.

3.2 The bisection step

In this section, referring to Algorithm 1 we briefly discuss, how the bisection step is executed. We therefore introduce the following definitions.

Definition 3.1. *Let $y \in \mathbb{R}$ be a real number. We define*

$$\lfloor y \rfloor := \max\{c \in \mathbb{Z} \mid c \leqslant y\},$$
$$\lceil y \rceil := \min\{c \in \mathbb{Z} \mid y \leqslant c\} \text{ and}$$
$$[y] := \begin{cases} \lceil y \rceil & \text{if } y + 0.5 \geqslant \lceil y \rceil \\ \lfloor y \rfloor & \text{else} \end{cases}.$$

If $y = (y_1, ..., y_n)^\top \in \mathbb{R}^n$ is a vector, we define

$$\lfloor y \rfloor := (\lfloor y_1 \rfloor, ..., \lfloor y_n \rfloor)^\top,$$
$$\lceil y \rceil := (\lceil y_1 \rceil, ..., \lceil y_n \rceil)^\top \text{ and}$$
$$[y] := ([y_1], ..., [y_n])^\top.$$

Now, let Assumption 2.9 be fulfilled. We describe the bisection step for a given subbox $\tilde{B} = [\underline{b'}, \overline{b'}] = \tilde{X} \times \tilde{Y} = [\underline{x'}, \overline{x'}] \times [\underline{y'}, \overline{y'}] \subseteq B$.

- Determine the largest Index $j \in \operatorname{argmax}\{(\overline{b}'_i - \underline{b}'_i) \mid i \in \{1, ..., r\}\}$, which is given by $j = \max\{\operatorname{argmax}\{(\overline{b}'_i - \underline{b}'_i) \mid i \in \{1, ..., r\}\}\}$.

- If $j \in \{m + 1, ..., r\}$, set

$$\tilde{B}^1 := [\underline{b}', (\overline{b}'_1, ..., \overline{b}'_{j-1}, \lfloor \frac{\underline{b}'_j + \overline{b}'_j}{2} \rfloor, \overline{b}'_{j+1}, ..., \overline{b}'_r)] \text{ and}$$

$$\tilde{B}^2 := [(\underline{b}'_1, ..., \underline{b}'_{j-1}, \lceil \frac{\underline{b}'_j + \overline{b}'_j}{2} \rceil, \underline{b}'_{j+1}, ..., \underline{b}'_r), \overline{b}'],$$

if $\underline{b}'_j + \overline{b}'_j$ is odd and

$$\tilde{B}^1 := [\underline{b}', (\overline{b}'_1, ..., \overline{b}'_{j-1}, \frac{\underline{b}'_j + \overline{b}'_j}{2}, \overline{b}'_{j+1}, ..., \overline{b}'_r)^\top] \text{ and}$$

$$\tilde{B}^2 := [(\underline{b}'_1, ..., \underline{b}'_{j-1}, \frac{\underline{b}'_j + \overline{b}'_j}{2} + 1, \underline{b}'_{j+1}, ..., \underline{b}'_r)^\top, \overline{b}'],$$

if $\underline{b}'_j + \overline{b}'_j$ is even.

- Else, set

$$\tilde{B}^1 := [\underline{b}', (\overline{b}'_1, ..., \overline{b}'_{j-1}, \frac{\underline{b}'_j + \overline{b}'_j}{2}, \overline{b}'_{j+1}, ..., \overline{b}'_r)^\top] \text{ and}$$

$$\tilde{B}^2 := [(\underline{b}'_1, ..., \underline{b}'_{j-1}, \frac{\underline{b}'_j + \overline{b}'_j}{2}, \underline{b}'_{j+1}, ..., \underline{b}'_r)^\top, \overline{b}'].$$

Note that in the bisection step, we choose the biggest index $j \in \operatorname{argmax}\{(\overline{b}'_i - \underline{b}'_i) \mid i \in \{1, ..., r\}\}$ for bisecting the box. In this way, we split the box along a longest side and additionally at an integer variable, if possible. In this case, the points in $\tilde{B} \backslash (\tilde{B}^1 \cup \tilde{B}^2)$ will not be considered further.

Moreover, we assumed $\underline{y}_i, \overline{y}_i \in \mathbb{Z}$ for all $i \in \{1, ..., n\}$ in Assumption 2.9.

Furthermore, for all subboxes $\tilde{B}$ obtained by using this bisection step it obviously holds $\underline{y}'_i, \overline{y}'_i \in \mathbb{Z}$ for all $i \in \{1, ..., n\}$. Therefore, we are able to check whether $\underline{b}'_j + \overline{b}'_j$ is odd or even in the case $j \in \{m + 1, ..., r\}$, since $\underline{b}'_j, \overline{b}'_j \in \mathbb{Z}$ holds then.

Also note that in the case $j \in \{m + 1, ..., r\}$ it generally does not hold $\tilde{B}^1 \cup \tilde{B}^2 = \tilde{B}$. This is intended, since obviously no $b \in \tilde{B} \backslash (\tilde{B}^1 \cup \tilde{B}^2)$ fulfills $b \in \tilde{B}^{\mathbb{Z}}$. Therefore, we do not 'loose' any feasible points for (MOMICP) due to our bisection method. The underlying idea is that this might save some computational time and leads to the following lemma and corollary.

Lemma 3.2. *Let Assumption 2.9 be fulfilled and let the subbox $\tilde{B} \in \mathbb{R}^r$ of B be derived from bisection steps. Then, it holds*

$$\tilde{B}^{g,\mathbb{Z}} \subseteq \tilde{B}^1 \cup \tilde{B}^2.$$

Corollary 3.3. *Let Assumption 2.9 be fulfilled and let the subbox $\tilde{B} \in \mathbb{R}^r$ of B be derived from bisection steps. Then the set of efficient points for (MOMICP) in $\tilde{B}$ is a subset of $\tilde{B}^1 \cup \tilde{B}^2$.*

The following example illustrates the bisection step.

Example 3.4. *We consider the boxes $X = [0, 4]$, $Y = [0, 8]$ and $B = X \times Y$. When bisecting B we obtain $j = \max\{\text{argmax}\{((4, 8)_i^\top - (0, 0)_i^\top) \mid i \in \{1, 2\}\}\} = 2$ and hence, $B^1 = [0, 4] \times [0, 4]$ and $B^2 = [0, 4] \times [5, 8]$. If we then bisect B^2, we obtain $j = \max\{\text{argmax}\{((4, 8)_i^\top - (0, 5)_i^\top) \mid i \in \{1, 2\}\}\} = 1$ and hence, $(B^2)^1 = [0, 2] \times [5, 8]$ and $(B^2)^2 = [2, 4] \times [5, 8]$. When applying another bisection step to $(B^2)^1$ we obtain $j = \max\{\text{argmax}\{((2, 8)_i^\top - (0, 5)_i^\top) \mid i \in \{1, 2\}\}\} = 2$ and hence, $((B^2)^1)^1 = [0, 2] \times [5, 6]$ and $((B^2)^1)^2 = [0, 2] \times [7, 8]$. The resulting boxes that we could bisect from there on are illustrated in the* *following figure. The dotted area illustrates the points that we discard due to the bisection step.*

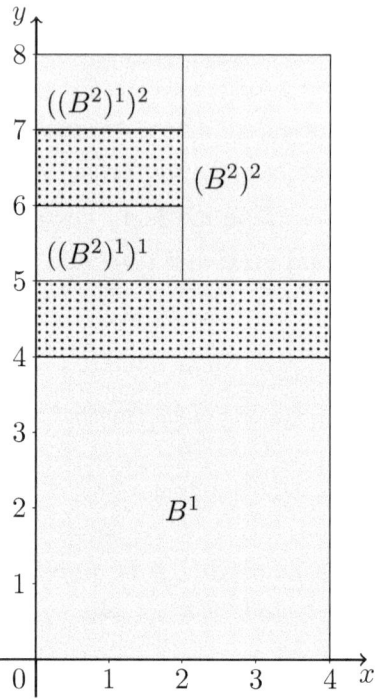

Figure 3.1: some bisection steps

3.3 A necessary feasibility condition

Referring to Algorithm 1, we introduce a necessary condition that a given subbox $\tilde{B}$ of B contains a feasible point for (MOMICP). Furthermore, we show that this condition is also sufficient in a special case. The idea is, to solve the optimization problem

$$\min \gamma$$
$$\text{s.t. } g(x,y) \leqslant \gamma \cdot e^q$$
$$(x,y) \in \tilde{B} \qquad\qquad (\text{FOP}(\tilde{B}))$$
$$\gamma \in \Gamma,$$

where $e^q := (1, ..., 1)^\top \in \mathbb{R}^q$ and Γ is the interval $\Gamma := [\underline{\gamma}, \overline{\gamma}]$ with

$$\underline{\gamma} := \min_{i \in \{1, ..., q\}} \min_{(x,y) \in B} g_i(x,y)$$

and

$$\overline{\gamma} := \max_{i \in \{1, ..., q\}} \max_{(x,y) \in B} g_i(x,y).$$

Since B is a nonempty and compact set and g_i is continuous for all $i \in \{1, ..., q\}$, the upper minima and maxima are well defined. We observe that the inequalities

$$\begin{aligned}
\underline{\gamma} &= \min_{i \in \{1, ..., q\}} \min_{(x,y) \in B} g_i(x,y) \\
&\leqslant \min_{(x,y) \in B} g_j(x,y) \\
&\leqslant g_j(x', y') \\
&\leqslant \max_{(x,y) \in B} g_j(x,y) \\
&\leqslant \max_{i \in \{1, ..., q\}} \max_{(x,y) \in B} g_i(x,y) = \overline{\gamma}.
\end{aligned} \qquad (3.1)$$

hold for all $(x', y') \in B$ and all $j \in \{1, ..., q\}$. Hence, for every $(x', y') \in B$ exists a $\gamma' \in \Gamma$ with $g(x', y') \leqslant \gamma' \cdot e^q$. Therefore, for every subbox $\tilde{B}$ of B and every $(x', y') \in \tilde{B}$ there is a $\gamma' \in \Gamma$ so that (x', y', γ') is feasible for $(\mathrm{FOP}(\tilde{B}))$. Hence, $(\mathrm{FOP}(\tilde{B}))$ is an optimization problem with linear objective function, convex and twice differentiable constraint functions on a nonempty and compact feasible set. Therefore the minimum value γ^* exists and can easily be computed. We observe that $\underline{\gamma} = \inf(\Gamma)$ can easily be obtained by solving q scalar valued convex optimization problems. However, $\overline{\gamma} = \sup(\Gamma)$ cannot be obtained easily, but since we are minimizing over $\gamma \in \Gamma$ for $(\mathrm{FOP}(\tilde{B}))$ and because of (3.1), it is equivalent to minimize over $\gamma \geqslant \underline{\gamma}$ instead of $\gamma \in \Gamma$. This eliminates the necessity to compute $\overline{\gamma}$ in order to solve $(\mathrm{FOP}(\tilde{B}))$.

The optimization problem $(\mathrm{FOP}(\tilde{B}))$ is an often used tool in optimization with convex constraints for finding feasible points. In the following theorem we use the minimum value γ^* of $(\mathrm{FOP}(\tilde{B}))$ in order to formulate a necessary and sufficient condition for a subbox $\tilde{B}$ of B to contain no feasible point for $(\mathrm{ROP}(\tilde{B}))$ or (MOMICP).

Theorem 3.5. *Let Assumption 2.9 be fulfilled and let $\tilde{B} = \tilde{X} \times \tilde{Y} \subseteq B$ be a subbox and γ^* the minimum value of $(\mathrm{FOP}(\tilde{B}))$.*

(i) It holds

$$\gamma^* > 0 \Leftrightarrow \tilde{B}^g = \varnothing \Rightarrow \tilde{B}^{g,\mathbb{Z}} = \varnothing.$$

(ii) If $\tilde{Y} = \{\tilde{y}\}$ for some $\tilde{y} \in Y \cap \mathbb{Z}^n$ (i.e. the integer variables in $\tilde{B}$ are fixed), it holds $\tilde{B}^g = \tilde{B}^{g,\mathbb{Z}}$ and therefore

$$\gamma^* > 0 \Leftrightarrow \tilde{B}^g = \varnothing \Leftrightarrow \tilde{B}^{g,\mathbb{Z}} = \varnothing.$$

Proof. For the proof of (i), we observe that obviously $\tilde{B}^g = \varnothing \Rightarrow \tilde{B}^{g,\mathbb{Z}} = \varnothing$ holds, because $\tilde{B}^{g,\mathbb{Z}} \subseteq \tilde{B}^g$. Now, consider a minimal solution $(x^*, y^*, \gamma^*) \in \tilde{X} \times \tilde{Y} \times \Gamma$ of $(\mathrm{FOP}(\tilde{B}))$.

For the first case, suppose $\gamma^* > 0$. Then for all $(x, y) \in \tilde{B}$ there is a $j \in \{1, ..., q\}$ with $g_j(x, y) \geqslant \gamma^* > 0$. Else, a contradiction to the minimality of γ^* would occur. Therefore, $g(x, y) \leqslant 0_q$ holds for no $(x, y) \in \tilde{B}$ and hence, $\tilde{B}^g = \varnothing$.

If, on the other hand, $\gamma^* \leqslant 0$ holds, then it is $(x^*, y^*) \in \tilde{B}$ with $g(x^*, y^*) \leqslant \gamma^* \cdot e^q \leqslant 0 \cdot e^q = 0_q$. Hence, $(x^*, y^*) \in \tilde{B}^g$ holds and $\tilde{B}^g \neq \varnothing$.

For the proof of (ii), assume that $\tilde{Y} = \{\tilde{y}\}$ for some $\tilde{y} \in Y \cap \mathbb{Z}^n$. Then it is easy to see that $\tilde{B}^{g,\mathbb{Z}} = \tilde{B}^g$ holds. Using (i) we are done. $\square$

In order to formulate a necessary feasibility condition we use the following corollary.

Corollary 3.6. *Let Assumption 2.9 be fulfilled and consider a subbox $\tilde{B}$ of B. Furthermore let γ^* be the minimal value of $(\mathrm{FOP}(\tilde{B}))$. Then it holds*

$$\tilde{B}^{g,\mathbb{Z}} \neq \varnothing \Rightarrow \gamma^* \leqslant 0.$$

Therefore, in line 7 of Algorithm 1, we check whether $\tilde{B}^j$ does not fulfill the necessary feasibility condition $\gamma^* \leqslant 0$.

3.4 Determining lower bounds

In this section, considering (MOMICP) we introduce a method that allows us to determine lower bounds of f on the set $\tilde{B}^{g,\mathbb{Z}}$ for a given subbox $\tilde{B}$ of B. In detail, we determine a set $L \subseteq \mathbb{R}^p$ of 'simple' structure that fulfills $f(\tilde{B}^{g,\mathbb{Z}}) \subseteq L + \mathbb{R}^p_+$. In order to achieve this, we determine lower bounds of f on $\tilde{B}^g$. Using $\tilde{B}^{g,\mathbb{Z}} \subseteq \tilde{B}^g$ the Lemma 3.7 follows immediately.

Lemma 3.7. *Consider (MOMICP) and a given subbox $\tilde{B}$ of B. Then it holds $f(\tilde{B}^{g,\mathbb{Z}}) \subseteq f(\tilde{B}^g)$.*

Now, we can observe that it is sufficient to determine a set $L \subseteq \mathbb{R}^p$ with $f(\tilde{B}^g) + \mathbb{R}^p_+ \subseteq L + \mathbb{R}^p_+$, because this then implies $f(\tilde{B}^{g,\mathbb{Z}}) \subseteq f(\tilde{B}^g) + \mathbb{R}^p_+ \subseteq L + \mathbb{R}^p_+$. In order to determine such lower bounds, we use the idea of Benson's Algorithm [3], which is connected with the concept of supporting hyperplanes.

Definition 3.8. *Let $Z \subseteq \mathbb{R}^p$ be a nonempty set, $\hat{z} \in \partial Z$ and $\lambda \in \mathbb{R}^p \backslash \{0_p\}$, where ∂Z is the boundary of the set Z. The Hyperplane*

$$H^{\lambda,\hat{z}} := \{h \in \mathbb{R}^p \mid \lambda^\top h = \lambda^\top \hat{z}\}$$

is called supporting Hyperplane (of Z), if $\lambda^\top z \geqslant \lambda^\top \hat{z}$ holds for all $z \in Z$.

We can obtain supporting hyperplanes for $f(\tilde{B}^g)$ by solving the optimization problem

$$
\begin{aligned}
\min \ & t \\
\text{s.t.} \ & f(b) \leqslant z + t \cdot e^p \\
& b \in \tilde{B}^g \\
& t \in \mathbb{R}
\end{aligned}
\qquad (\text{HPOP}_z(\tilde{B}))
$$

for elements $z \in \mathbb{R}^p$.

Assumption 3.9. *In addition to Assumption 2.9, we assume regularity for the optimization problem $(\text{HPOP}_z(\tilde{B}))$ for any $z \in \mathbb{R}^p$ for which we consider $(\text{HPOP}_z(\tilde{B}))$. This means any solution of $(\text{HPOP}_z(\tilde{B}))$ also induces a KKT point of $(\text{HPOP}_z(\tilde{B}))$.*

The following lemma allows us to construct supporting hyperplanes of $f(\tilde{B}^g)$ for a subbox $\tilde{B}$ of B.

Lemma 3.10. *Consider (MOMICP), a subbox $\tilde{B}$ of B and an element $\hat{z} \in \mathbb{R}^p$ and let Assumption 3.9 for $\hat{z}$ be fulfilled. If $\tilde{B}^g$ is nonempty, then the following statements hold.*

(i) *The optimization problem* $(\mathrm{HPOP}_{\hat{z}}(\tilde{B}))$ *has a solution* $(\hat{b}, \hat{t}) \in \tilde{B}^g \times \mathbb{R}$ *with Lagrange multiplier $\hat{\lambda} \in \mathbb{R}^p$ for the constraint* $f(b) \leqslant \hat{z} + t \cdot e^p$.

(ii) *The hyperplane $H^{\hat{\lambda}, \hat{p}(\hat{z})}$ with $\hat{p}(\hat{z}) := \hat{z} + \hat{t} \cdot e$ is a supporting hyperplane of $f(\tilde{B}^g)$ and it holds* $f(\tilde{B}^g) + \mathbb{R}^p_+ \subseteq H^{\hat{\lambda}, \hat{p}(\hat{z})} + \mathbb{R}^p_+$.

In line 10 of Algorithm 1 we solve $(\mathrm{HPOP}_{\hat{z}}(\tilde{B}))$ for $\tilde{B} = \tilde{B}^j$ and certain elements $\hat{z} \in Z$ of a given nonempty and finite set $Z \subseteq \mathbb{R}^p$. We consider the set Z as given for now. In Section 3.5 we will go in detail regarding the choice for the set Z.

We obtain supporting hyperplanes $H^{\hat{\lambda}, \hat{p}(\hat{z})}$ with $f(\tilde{B}^g) + \mathbb{R}^p_+ \subseteq H^{\hat{\lambda}, \hat{p}(\hat{z})} + \mathbb{R}^p_+$ for all $\hat{z} \in Z$. Then we define

$$L := \partial\left(\bigcap_{\hat{z} \in Z}(H^{\hat{\lambda}, \hat{p}(\hat{z})} + \mathbb{R}^p_+)\right),$$

i.e. as the boundary of the polyhedron $\bigcap_{\hat{z} \in Z}(H^{\hat{\lambda}, \hat{p}(\hat{z})} + \mathbb{R}^p_+)$. According to Lemma 3.10, it holds

$$f(\tilde{B}^{g,\mathbb{Z}}) \subseteq f(\tilde{B}^g) + \mathbb{R}^p_+ \subseteq \bigcap_{\hat{z} \in Z}(H^{\hat{\lambda}, \hat{p}(\hat{z})} + \mathbb{R}^p_+) = L + \mathbb{R}^p_+.$$

3.5 Determining upper bounds

In the previous section, considering (MOMICP) we investigated how to derive lower bounds of f on $\tilde{B}^{g,\mathbb{Z}}$ for a given subbox $\tilde{B}$ of B. In this section, we introduce a method to obtain upper bounds of f on $B^{g,\mathbb{Z}}$. Later, we use these upper bounds and compare them with the lower bounds of f on $\tilde{B}^{g,\mathbb{Z}}$. We introduce the concept of local upper bounds for this. This comparison will set the basis for us to check, whether $\tilde{B}$ can contain a efficient point for (MOMICP). This information will be used for our discarding test.

In order to derive upper bounds of f on $B^{g,\mathbb{Z}}$, we try to find feasible points $b \in \tilde{B}^{g,\mathbb{Z}}$ and evaluate their respective function values. However, finding such points is a difficult task and in general, there are no methods known to achieve this within an adequate amount of time. There are however heuristic approaches like the feasibility pump [4]. Additionally, there is the concept of granularity, which would need additional assumptions, however.

For the sake of simplicity, we use a heuristic approach that simply rounds points $b = (x, y) \in \tilde{B}^g$ in order to match the constraint $y \in \mathbb{Z}^n$. Afterwards, we check whether the obtained point still fulfills the constraints given by g.

Throughout proceeding Algorithm 1 we obtain points $b \in \tilde{B}^g$. For example, recall the necessary feasibility condition. After solving (FOP($\tilde{B}$)) for a subbox $\tilde{B}$ of B we obtain a minimal solution (x^*, y^*, γ^*). If $\tilde{B}$ fulfills the necessary feasibility condition $\gamma^* \leqslant 0$, it holds $b^* := (x^*, y^*) \in \tilde{B}^g$ (Theorem 3.5).

Additionally, after solving (HPOP$_z$($\tilde{B}$)) we obtain a minimal solution $(b^*, t^*) \in \tilde{B}^g \times \mathbb{R}$.

We are then able to apply the rounding step to such points b^* and try to obtain a point $b \in \tilde{B}^{g,\mathbb{Z}}$ in that way.

At first, we introduce the following definition.

Definition 3.11. *Let Assumption 2.9 be fulfilled and consider a subbox $\tilde{B}$ of B. For an element $b = (x, y) \in \tilde{B}$ we define*

$$b^{\mathbb{Z}} := (x, [y]) = (x, [y_1], ..., [y_n]).$$

Lemma 3.12. *Let Assumption 2.9 be fulfilled and consider a subbox $\tilde{B} = [\underline{x}', \overline{x}'] \times [\underline{y}', \overline{y}'] \subseteq B$ obtained by bisection steps starting with B. Let b be an element $b = (x, y) \in \tilde{B}$. It holds*

$$b^{\mathbb{Z}} \in \tilde{B}^{\mathbb{Z}}.$$

Proof. Since we obtained $\tilde{B}$ by bisection steps, it holds $\underline{y}'_i, \overline{y}'_i \in \mathbb{Z}$ for all $i \in \{1, ..., n\}$. Furthermore, it holds $\underline{x}' \leqslant x \leqslant \overline{x}'$ and $\underline{y}' \leqslant y \leqslant \overline{y}'$, because $(x, y) \in \tilde{B}$. This implies

$$\underline{y}' = [\underline{y}'] \leqslant [y] \leqslant [\overline{y}'] = \overline{y}'.$$

Therefore, $b^{\mathbb{Z}} = (x, [y]) \in \tilde{B}$ holds and since $[y] \in \mathbb{Z}^n$, it is $b^{\mathbb{Z}} \in \tilde{B}^{\mathbb{Z}}$. $\square$

Summarizing, for any element $b = (x, y) \in \tilde{B}^g \subseteq \tilde{B}$ it holds $b^{\mathbb{Z}} \in \tilde{B}^{\mathbb{Z}}$. However, it not necessarily holds $g(b^{\mathbb{Z}}) \leqslant 0_q$. Therefore, we cannot assure $b^{\mathbb{Z}} \in \tilde{B}^{g,\mathbb{Z}}$, but it is easy to verify whether this holds.

Moreover, we are interested in finding many points $b \in \tilde{B}^{g,\mathbb{Z}}$ with function values that are nondominated by other known function values of feasible points. Therefore, we introduce the set $\mathcal{L}_{\mathcal{PNS}}$. At the beginning of Algorithm 1, we initialize $\mathcal{L}_{\mathcal{PNS}} = \emptyset$. Then, we handle $\mathcal{L}_{\mathcal{PNS}}$ as follows.

When considering a subbox $\tilde{B}$ of B we obtain a point $b \in \tilde{B} \subseteq B$ after solving any of the optimization problems $(\text{ROP}(\tilde{B})_i)$ for $i \in \{1, ..., p\}$, $(\text{FOP}(\tilde{B}))$ and $(\text{HPOP}_z(\tilde{B}))$ for some $z \in \mathbb{R}^p$ and follow these steps.

Algorithm 2 Updating $\mathcal{L}_{\mathcal{PNS}}$

INPUT: $b \in B$, $\mathcal{L}_{\mathcal{PNS}}$
OUTPUT: $\mathcal{L}_{\mathcal{PNS}}$

1: Calculate $b^{\mathbb{Z}}$.
2: **if** $b^{\mathbb{Z}} \in B^{g,\mathbb{Z}}$ **then**
3:   Calculate $w := f(b^{\mathbb{Z}})$.
4:   **if** $\nexists w' \in \mathcal{L}_{\mathcal{PNS}} : w' \leq w$ **then**
5:     Set $\mathcal{L}_{\mathcal{PNS}} = \mathcal{L}_{\mathcal{PNS}} \cup \{w\}$.
6:     Set $\mathcal{L}_{\mathcal{PNS}} = \mathcal{L}_{\mathcal{PNS}} \backslash \{w' \in \mathcal{L}_{\mathcal{PNS}} \mid w \lneq w'\}$.
7:   **end if**
8: **end if**

We observe that $\mathcal{L}_{\mathcal{PNS}}$ is a finite and stable set that contains function values of points in $B^{g,\mathbb{Z}}$. We use these points in order to derive so called local upper bounds which we introduce in the following definition.

Definition 3.13. *[12] Let Assumption 2.9 be fulfilled and let $\mathcal{N} \subseteq \mathbb{R}^p$ be a finite and stable set of points and $\hat{Z} \in \mathbb{R}^p$ be a box with $f(B) \subseteq$ int$(\hat{Z})$. We define the search region*

$$S := \{z \in \text{int}(\hat{Z}) \mid w \nleq z \text{ for all } w \in \mathcal{N}\}.$$

Furthermore, a list $\mathcal{L} \subseteq \hat{Z}$ is called a local upper bound set with respect to $\mathcal{N}$, if

(i) $\forall z \in S \, \exists w \in \mathcal{L} : z < w$,

(ii) $\forall z \in \text{int}(\hat{Z}) \backslash S \, \forall w \in \mathcal{L} : z \nless w$ *and*

(iii) $\forall w^1, w^2 \in \mathcal{L} : w^1 \nless w^2$ *or* $w^1 = w^2$.

Remark 3.14. *Here, we consider the set* $\hat{Z} \in \mathbb{R}^p$ *with* $f(B) \subseteq \text{int}(\hat{Z})$ *as given. One could determine such a box by using interval arithmetic [15].*

We want to make use of a local upper bound set with respect to $\mathcal{L}_{\mathcal{PNS}}$. Therefore, we introduce the set $\mathcal{L}_{\mathcal{LUB}}$. We initialize $\mathcal{L}_{\mathcal{LUB}} = \{\sup(\hat{Z})\}$ at the beginning of Algorithm 1. Then, we have to update $\mathcal{L}_{\mathcal{LUB}}$ everytime after Algorithm 2 has been executed and the respective $b^{\mathbb{Z}}$ was added to $\mathcal{L}_{\mathcal{PNS}}$. Therefore, we execute Algorithm 3 from [12]. This algorithm assures that $\mathcal{L}_{\mathcal{LUB}}$ is a local upper bound set w.r.t. the finite and stable set $\mathcal{L}_{\mathcal{PNS}}$.

The following figure illustrates the local upper bound set $\mathcal{L}_{\mathcal{LUB}}$ with respect to the set $\mathcal{L}_{\mathcal{PNS}}$ for the case $p = 2$. Hereby, it is $\overline{M}_1 := \sup(\hat{Z})_1$ and $\overline{M}_2 := \sup(\hat{Z})_2$.

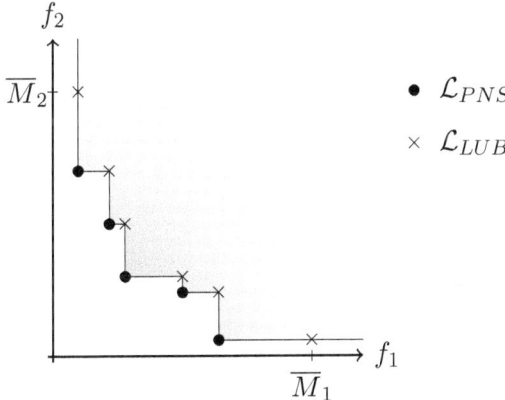

Figure 3.2: $\mathcal{L}_{\mathcal{PNS}}$ and $\mathcal{L}_{\mathcal{LUB}}$ cf. [12], Figure 1

We will use the concept of local upper bounds to introduce the discarding test in the next section.

3.6 The discarding test and termination rule

In this section, under Assumption 2.9 we use the method for determining lower bounds of f on $\tilde{B}^{g,\mathbb{Z}}$ for a given subbox $\tilde{B}$ of B from Section 3.4 as well as the concept of local upper bounds from Section 3.5 in order to introduce a discarding test. The following lemma is taken from [16] and will be used for proving the fundamental theorem for formulating our discarding test, which is inspired by the discarding test proposed in [16].

Lemma 3.15. *Let Assumption 2.9 be fulfilled and let $\mathcal{L}$ be a local upper bound set with respect to a finite and stable set of points $\mathcal{N} \subseteq f(B^{g,\mathbb{Z}})$. For every $\overline{z} \in \mathcal{N}$ and for every $j \in \{1,...,p\}$ there is a $z \in \mathcal{L}$ with $\overline{z}_j = z_j$ and $\overline{z}_i < z_i$ for all $i \in \{1,...,p\}\backslash\{j\}$.*

The following theorem will set the basis for the discarding test.

Theorem 3.16. *Let Assumption 2.9 be fulfilled and consider a subbox $\tilde{B}$ of B. Let $\mathcal{L}_{\mathcal{PNS}} \subseteq f(B^{g,\mathbb{Z}})$ be a finite and stable set. Moreover, let $\mathcal{L}_{\mathcal{LUB}}$ be the local upper bound set w.r.t. $\mathcal{L}_{\mathcal{PNS}}$ as described in Section 3.5. If*

$$\forall z \in \mathcal{L}_{\mathcal{LUB}} : z \notin f(\tilde{B}^g) + \mathbb{R}^p_+ \tag{3.2}$$

holds, then $\tilde{B}$ does not contain any efficient point for (MOMICP).

Proof. Assume that there is some efficient point $x^* \in \tilde{B}$ for (MOMICP). This implies $x^* \in \tilde{B}^{g,\mathbb{Z}}$. Since $f(\tilde{B}^{g,\mathbb{Z}}) + \mathbb{R}^p_+ \subseteq f(\tilde{B}^g) + \mathbb{R}^p_+$ by Lemma 3.7, it holds $z \notin f(\tilde{B}^{g,\mathbb{Z}}) + \mathbb{R}^p_+$ for all $z \in \mathcal{L}_{\mathcal{LUB}}$. This is implies

$$f(x^*) \nleq z \text{ for all } z \in \mathcal{L}_{\mathcal{LUB}}. \tag{3.3}$$

Because $\mathcal{L}_{\mathcal{LUB}}$ is a local upper bound set w.r.t. $\mathcal{L}_{\mathcal{PNS}}$ we conclude with Definition 3.13 (i) $f(x^*) \notin S$. This implies that there exists a point $q \in \mathcal{L}_{\mathcal{PNS}}$ with $q \leqslant f(x^*)$. Since $q \in \mathcal{L}_{\mathcal{PNS}}$, it is $q = f(x')$ for some $x' \in B^{g,\mathbb{Z}}$. Because x^* is efficient for (MOMICP), it follows $q = f(x') = f(x^*)$. Lemma 3.15 implies that there is a point $z' \in \mathcal{L}_{\mathcal{CUB}}$ with $f(x^*) \leqslant z'$, which is a contradiction to (3.3). Thus, $\tilde{B}$ does not contain any efficient point for (MOMICP). $\qquad\square$

Condition (3.2) can be tested by solving $(\mathrm{HPOP}_z(\tilde{B}))$ for the respective local upper bounds $z \in \mathcal{L}_{\mathcal{CUB}}$ and the box $\tilde{B}$. However, in order to save some computational time, we do not simply solve $(\mathrm{HPOP}_z(\tilde{B}))$ for every $z \in \mathcal{L}_{\mathcal{CUB}}$. Instead, we use Lemma 3.10 to determine supporting hyperplanes $H^{\hat{\lambda},\hat{p}}$ of $f(\tilde{B}^g) + \mathbb{R}^p_+$ after we solved $(\mathrm{HPOP}_z(\tilde{B}))$ for any $z \in \mathcal{L}_{\mathcal{CUB}}$. Before solving $(\mathrm{HPOP}_{z'}(\tilde{B}))$ for the next local upper bound $z' \in \mathcal{L}_{\mathcal{CUB}}$, we check whether $\hat{\lambda}^\top z' \geqslant \hat{\lambda}^\top \hat{p}$ holds for all determined hyperplanes. This is necessary for $z' \in f(\tilde{B}^g) + \mathbb{R}^p_+$ and only then we have to solve $(\mathrm{HPOP}_{z'}(\tilde{B}))$.

We summarize some theoretical results on $(\mathrm{HPOP}_z(\tilde{B}))$ and how we use them. Then, we formulate the discarding test in detail. Therefore, we assume that we have a given accuracy parameter $\varepsilon > 0$, a local upper bound $z \in \mathcal{L}_{\mathcal{CUB}}$ and the solution $(b^*, t^*) \in \tilde{B}^g \times \mathbb{R}$ of $(\mathrm{HPOP}_z(\tilde{B}))$.

(1) If $t^* \leqslant 0$ holds, then it obviously is $z \in f(\tilde{B}^g) + \mathbb{R}^p_+$. Hence, we cannot apply Theorem 3.16 and $\tilde{B}$ may contain efficient points for (MOMICP). Thus, $\tilde{B}$ cannot be discarded and we distinguish between the following two subcases.

   (a) If $t^* < -\frac{\varepsilon}{2}$ holds, then we stop the discarding test and bisect $\tilde{B}$ later.

   (b) If $-\frac{\varepsilon}{2} \leqslant t^* \leqslant 0$ holds, then we construct a supporting hyperplane using Lemma 3.10 in order to improve the outer approximation of $f(\tilde{B}^g) + \mathbb{R}^p_+$ and consider the next local upper bound.

(2) If $t^* > 0$ holds, then it obviously is $z \notin f(\tilde{B}^g) + \mathbb{R}^p_+$. We then construct a supporting hyperplane using Lemma 3.10 in order to improve the outer approximation of $f(\tilde{B}^g) + \mathbb{R}^p_+$ and consider the next local upper bound.

Thereby we apply the strategy from [16], which is motivated and explained in more detail also in [16].

The following algorithm describes the discarding test in detail. Therefore, we introduce a flag $\mathcal{D}$ for discarding a box and a flag $\mathcal{B}$ for bisecting a box. In addition, we introduce a flag $\mathcal{F}$ that states, whether the lists $\mathcal{L}_{\mathcal{PNS}}$ and $\mathcal{L}_{\mathcal{LUB}}$ are fixed, i.e. no further points will be added to them. For the basic algorithm, we set $\mathcal{F} = \texttt{false}$ for now. Furthermore, let u^i be the i-th unit vector in $\mathbb{R}^p$ for $i \in \{1, ..., p\}$.

Algorithm 3 Discarding test

INPUT: (MOMICP), a subbox $\tilde{B}$ of B, $\mathcal{L}_{\mathcal{PNS}}$, $\mathcal{L}_{\mathcal{CUB}}$, $\varepsilon > 0$, $\mathcal{F}$

OUTPUT: $\mathcal{D}$, $\mathcal{B}$, $\mathcal{L}_{\mathcal{PNS}}$, $\mathcal{L}_{\mathcal{CUB}}$

1: Set $\mathcal{L}_{\mathcal{PNS}}{}^* \leftarrow \mathcal{L}_{\mathcal{PNS}}$, $\mathcal{L}_{\mathcal{CUB}}{}^* \leftarrow \mathcal{L}_{\mathcal{CUB}}$.
2: Solve $(\mathrm{ROP}(\tilde{B})_i)$ with solution $b^i \in \tilde{B}^g$ for all $i \in \{1, ..., p\}$.
3: **if** $\neg \mathcal{F}$ **then**
4:     Apply Algorithm 2 to update $\mathcal{L}_{\mathcal{PNS}}$ with b^i for $i \in \{1, ..., p\}$.
5:     Update $\mathcal{L}_{\mathcal{CUB}}$ using Algorithm 3 from [12].
6: **end if**
7: Set $a \leftarrow (f_1(b^1), ..., f_p(b^p))^\top$.
8: Set $\mathcal{H} \leftarrow \{H^{\mathbf{u}^i, a} \mid i \in \{1, ..., p\}\}$.
9: Set $\mathcal{D} \leftarrow \mathtt{true}$, $\mathcal{B} \leftarrow \mathtt{false}$.
10: **for** $z \in \mathcal{L}_{\mathcal{CUB}}{}^*$ **do**
11:     **if** $\forall H^{\lambda, z'} \in \mathcal{H} : \lambda^\top z \geq \lambda^\top z'$ **then**
12:        Solve $(\mathrm{HPOP}_z(\tilde{B}))$ with solution $(b^*, t^*) \in \tilde{B} \times \mathbb{R}$ and Lagrange multiplier $\quad \lambda \in \mathbb{R}^p$ for the constraint $f(b) \leq z + t \cdot e^p$.
13:        **if** $\neg \mathcal{F}$ **then**
14:           Apply Algorithm 2 to update $\mathcal{L}_{\mathcal{PNS}}$ with b^*.
15:           Update $\mathcal{L}_{\mathcal{CUB}}$ using Algorithm 3 from [12].
16:        **end if**
17:        **if** $t^* < -\frac{\varepsilon}{2}$ **then**
18:           Set $\mathcal{D} \leftarrow \mathtt{false}$, $\mathcal{B} \leftarrow \mathtt{true}$.
19:           **break for** − **loop**
20:        **else**
21:           Set $\mathcal{H} \leftarrow \mathcal{H} \cup \{H^{\hat{\lambda}, z + t^* \cdot e^p}\}$.
22:           **if** $t^* \leq 0$ **then**
23:              Set $\mathcal{D} \leftarrow \mathtt{false}$.
24:           **end if**
25:        **end if**
26:     **end if**
27: **end for**

Referring to Algorithm 1, we discard a box $\tilde{B}$, if Algorithm 3 applied to it returns $\mathcal{D} = \texttt{true}$.

Our termination rule for the basic Branch-and-Bound algorithm for a given subbox $\tilde{B}$ is that Algorithm 3 returns $\mathcal{D} = \texttt{false}$ and $\mathcal{B} = \texttt{false}$. In this case, there is a local upper bound $z \in \mathcal{L}_{\mathcal{C}\mathcal{U}\mathcal{B}}$ with $z \in f(\tilde{B}^g) + \mathbb{R}_+^p$ and no local upper bound suggests a further bisection of $\tilde{B}$. This suggests that $f(\tilde{B}^{g,\mathbb{Z}})$ has reached a certain 'accuracy' in $f(B^{g,\mathbb{Z}})$ without getting discarded. Hence, this box is of interest for us, because it might contain efficient points for (MOMICP). Later, we enhance Algorithm 1 by further investigating the boxes on $\mathcal{L}_{\mathcal{S}}$. For example, we then assure a certain 'accuracy' in the pre-image space as well by bisecting these boxes until their box width is less or equal a given accuracy parameter δ.

However, we want to prove the exactness of Algorithm 3 at first. Therefore, we show that the discarding test returns $\mathcal{D} = \texttt{false}$ for every box that contains an efficient point for (MOMICP). Hence, we do not discard such boxes throughout proceeding Algorithm 1.

Theorem 3.17. *Let $\tilde{B}$ be a subbox of B that contains an efficient point $x^* \in \tilde{B}^{g,\mathbb{Z}}$ of (MOMICP). Applying Algorithm 3 then returns $\mathcal{D} = \texttt{false}$.*

Proof. Assume that Algorithm 3 returns $\mathcal{D} = \texttt{true}$. This implies that for all local upper bounds $z \in \mathcal{L}_{\mathcal{C}\mathcal{U}\mathcal{B}}$ it holds $t^* > 0$, where (b^*, t^*) is the solution of $(\text{HPOP}_z(\tilde{B}))$. Hence, $z \notin f(\tilde{B}^g) + \mathbb{R}_+^p$ holds for all $z \in \mathcal{L}_{\mathcal{C}\mathcal{U}\mathcal{B}}$. Theorem 3.16 then implies that $\tilde{B}$ does not contain any efficient point for (MOMICP). This is a contradiction to the assumption that $\tilde{B}$ contains the efficient point x^*. $\qquad\square$

We use this theorem to prove the exactness of Algorithm 1. This means after applying Algorithm 1 the set of efficient points E of

(MOMICP) is a subset of the union of all boxes of $\mathcal{L_S}^1$, or in other words $\mathcal{L_S}^1$ is a cover of E.

Theorem 3.18. *Let E be the efficient set of (MOMICP). After applying Algorithm 1 the set $\mathcal{L_S}^1$ is a cover of E.*

Proof. At first we observe that it holds $B = \bigcup \mathcal{L_S}^1 \cup \bigcup \mathcal{L_{NS}}^1 \cup \mathcal{BS}$, where $\mathcal{BS}$ is the set of all points that are discarded due to the bisection step. Then it obviously holds $E \subseteq \bigcup \mathcal{L_S}^1 \cup \bigcup \mathcal{L_{NS}}^1 \cup \mathcal{BS}$. Corollary 3.3 implies $E \nsubseteq \mathcal{BS}$ and Theorem 3.17 and Corollary 3.6 imply $E \nsubseteq \bigcup \mathcal{L_{NS}}^1$. Thus, $E \subseteq \bigcup \mathcal{L_S}^1$ holds. $\square$

4 Enhancing Algorithm 1

In this chapter, we introduce modifications that enhance the basic Branch-and-Bound algorithm for (MOMICP), we introduced in Chapter 3. We follow different goals with these modifications. We would like to reduce the amount of computational time, the algorithm requires, as well as provide a 'better' cover of the efficient set of (MOMICP). In order to create an understanding of the precision of a cover of the efficient set of (MOMICP), we introduce the following definition.

Definition 4.1. *Let $\mathcal{L}^1, \mathcal{L}^2 \subseteq \mathcal{P}(\mathbb{R}^r)$ be covers of the efficient set of (MOMICP). The cover $\mathcal{L}^1$ is called more precise than $\mathcal{L}^2$, if $\bigcup \mathcal{L}^1 \subseteq \bigcup \mathcal{L}^2$ holds.*

4.1 Preinitialization

The first idea is to compute some imagepoints of f on $B^{g,\mathbb{Z}}$ before starting the basic algorithm. Using these points, we then initialize $\mathcal{L}_{\mathcal{PNS}}$ and $\mathcal{L}_{\mathcal{CUB}}$.

Therefore, let imgp $\in \mathbb{N}$ be a given number. We want to distribute evenly many points on every possible combination of integer variables and then create an equidistant grid on the remaining continuous variables in a way that we obtain about imgp points in B. We then check for each of these points $b \in B$, if $b \in B^{g,\mathbb{Z}}$ holds and in this case

© Springer Fachmedien Wiesbaden GmbH, part of Springer Nature 2020
S. Rocktäschel, *A Branch-and-Bound Algorithm for Multiobjective Mixed-integer Convex Optimization*, BestMasters, https://doi.org/10.1007/978-3-658-29149-5_4

update $\mathcal{L}_{\mathcal{PNS}}$ as well as $\mathcal{L}_{\mathcal{CUB}}$.

The following algorithm describes this procedure in detail.

Algorithm 4 Preinitialization

INPUT: (MOMICP), imgp $\in \mathbb{N}$

OUTPUT: $\mathcal{L}_{\mathcal{PNS}}$, $\mathcal{L}_{\mathcal{CUB}}$

1: **if** imgp $= 0 \vee m = 0$ **then**
2:    Set $\mathcal{L}_{\mathcal{PNS}} \leftarrow \varnothing$, $\mathcal{L}_{\mathcal{CUB}} \leftarrow \{\sup(\hat{Z})\}$.
3: **else**
4:    Set grd $\leftarrow (\sum_{i=1}^{m} \frac{\overline{x}_i - \underline{x}_i}{m \cdot \text{imgp}^{\frac{1}{m}}}) \cdot (\prod_{i=1}^{n}(\overline{y}_i - \underline{y}_i + 1))^{\frac{1}{m}}$.
5:    Set Pos $\leftarrow 0 \cdot e^r$ and $\mathcal{FV} \leftarrow \varnothing$.
6:    **while** $\underline{x} + \text{grd} \cdot (\text{Pos}_1, ..., \text{Pos}_m)^\top \leqslant \overline{x} \wedge \underline{y} + (\text{Pos}_{m+1}, ..., \text{Pos}_r)^\top \leqslant \overline{y}$ **do**
7:       Set $b \leftarrow (\underline{x} + \text{grd} \cdot (\text{Pos}_1, ..., \text{Pos}_m)^\top, \underline{y} + (\text{Pos}_{m+1}, ..., \text{Pos}_r)^\top)$.
8:       **if** $b \in B^{g,\mathbb{Z}}$ **then**
9:          Set $\mathcal{FV} \leftarrow \mathcal{FV} \cup \{b\}$.
10:       **end if**
11:       Set $\text{Pos}_1 \leftarrow \text{Pos}_1 + 1$.
12:       **for** $i = 1, ..., r - 1$ **do**
13:          **if** $(i \leqslant m \wedge \underline{x}_i + \text{grd} \cdot \text{Pos}_i > \overline{x}_i) \vee (i > m \wedge \underline{y}_{i-m} + \text{Pos}_i > \overline{y}_{i-m})$ **then**
14:             Set $\text{Pos}_i \leftarrow 0$.
15:             Set $\text{Pos}_{i+1} \leftarrow \text{Pos}_{i+1} + 1$.
16:          **end if**
17:       **end for**
18:    **end while**
19:    Set $\mathcal{L}_{\mathcal{PNS}} \leftarrow \varnothing$, $\mathcal{L}_{\mathcal{CUB}} \leftarrow \{\sup(\hat{Z})\}$.
20:    **for** $b \in \mathcal{FV}$ **do**
21:       Update $\mathcal{L}_{\mathcal{PNS}}$ using Algorithm 2 with input b.
22:       Update $\mathcal{L}_{\mathcal{CUB}}$ using Algorithm 3 from [12].
23:    **end for**
24: **end if**

Remark 4.2. *Note that for* imgp > 0 *the preinitialization can only* *be applied, if $m > 0$ holds. Otherwise, the grid parameter* grd *is not defined. However, since we only have to consider integer variables for $m = 0$, there is only a finite number of feasible points anyways. Thus, a preinitialization might not be that useful regarding decreasing computational time. Hence, Algorithm 4 returns $\mathcal{L}_{\mathcal{PNS}} = \varnothing$ and $\mathcal{L}_{\mathcal{CUB}} = \varnothing$ in the case $m = 0$ (lines 1,2). Also note that we consider all $y \in Y \cap \mathbb{Z}^n$ for calculating points $b \in B$ during the preinitialization due to the method of using the position vector* Pos. *This implies that we consider all points $(\underline{x}, y)$ for $y \in Y \cap \mathbb{Z}^n$ and hence at least $|Y \cap \mathbb{Z}^n|$ points for* imgp > 0 *and $m > 0$ for the preinitialization. This might be much more than the intended number* imgp *of points that we want to consider. One should take this fact into account when using the preinitialization.*

In line 4, we calculate the grid parameter grd that is used for creating an equidistant grid on the continuous variables. The idea behind this formula is to consider roughly imgp points during the preinitialization. If there were no integer variables and $m = 1$, we would simply use $\frac{\overline{x}_1 - \underline{x}_1}{\text{imgp}}$ as grid parameter, what is also done in Algorithm 4 in that case. For $m > 1$ and $n = 0$ we take the mean value of interval lengths $\sum_{i=1}^{m} \frac{\overline{x}_i - \underline{x}_i}{m}$ into consideration and divide this by $\text{imgp}^{\frac{1}{m}}$ in order to obtain roughly $\text{imgp}^{\frac{1}{m}}$ grid points on each X_i for $i \in \{1, ..., m\}$ and hence $(\text{imgp}^{\frac{1}{m}})^m = \text{imgp}$ grid points on X. If we have to consider additional integer variables in the case $n > 0$, we want to distribute evenly many grid points on every possible $y \in Y \cap \mathbb{Z}^n$. There are $\prod_{i=1}^{n}(\overline{y}_i - \underline{y}_i + 1)$ possible $y \in Y \cap \mathbb{Z}^n$. Hence, we have to divide the number of grid points on X by that number in order to still obtain roughly imgp grid points in total on $X \times Y$. We achieve this by multiplying the grid parameter from the case $n = 0$, which is

$\sum_{i=1}^{m} \frac{\overline{x}_i - \underline{x}_i}{m \cdot \text{imgp}^{\frac{1}{m}}}$, with $(\prod_{i=1}^{n}(\overline{y}_i - \underline{y}_i + 1))^{\frac{1}{m}}$. By doing that, the amount

of grid points we consider on each X_i is now roughly $\frac{\text{imgp}^{\frac{1}{m}}}{(\prod_{i=1}^{n}(\overline{y}_i - \underline{y}_i + 1))^{\frac{1}{m}}}$.

Hence, we consider about $(\frac{\text{imgp}^{\frac{1}{m}}}{(\prod_{i=1}^{n}(\overline{y}_i - \underline{y}_i + 1))^{\frac{1}{m}}})^m = \frac{\text{imgp}}{\prod_{i=1}^{n}(\overline{y}_i - \underline{y}_i + 1)}$ grid

points on X and thus $\frac{\text{imgp}}{\prod_{i=1}^{n}(\overline{y}_i - \underline{y}_i + 1)} \cdot \prod_{i=1}^{n}(\overline{y}_i - \underline{y}_i + 1) = \text{imgp}$ grid

points on $X \times Y$. This leads to the formula for the grid parameter

grd that we use in line 4 of Algorithm 4.

In line 5, we initialize the position vector Pos and the set of obtained

function values $\mathcal{FV} \subseteq f(B^{g,\mathbb{Z}})$. In line 6, we start going through the

box B using the position vector Pos until we obtain a point that is

not in B anymore. In line 7, we calculate the current point that has

to be considered. In the first iteration, this will be $b = (\underline{x}, \underline{y})$ and from

there on we use the position vector Pos to go through B. In lines

8-10, we check whether the current point b is in $B^{g,\mathbb{Z}}$ and add it to

$\mathcal{FV}$ in that case. In lines 11-17, we update the position vector Pos to

determine the next point. Finally, in lines 19-23, we use the points

$b \in \mathcal{FV} \subseteq B^{g,\mathbb{Z}}$ in order to initialize $\mathcal{L}_{\mathcal{PNS}}$ and $\mathcal{L}_{\mathcal{CUB}}$.

We use the preinitialization in order to obtain some local upper

bounds and function values of f on $B^{g,\mathbb{Z}}$ before applying Algorithm 1.

In that way, we might be able to discard certain boxes or verify the

termination rule after less iterations than we would need otherwise.

On the other hand, we might have to solve more optimization problems

in each iteration to apply the discarding test for example, because we

have more local upper bounds to consider for that. However, this is

just a heuristic approach.

Summarizing, we presume that we need less iterations but more

computational time per iteration, if we use the preinitialization before applying Algorithm 1. Applying the enhanced algorithm to certain test instances, we examine this relationship more precisely in Chapter 5.

4.2 Elimination step

In this section, we introduce a method that we can apply after Algorithm 1 in order to obtain a more precise approximation of the efficient set E of (MOMICP) than $\mathcal{L_S}^1$. The idea is to consider all boxes $\tilde{B} \in \mathcal{L_S}^1$ and check, if $\tilde{B}$ can be discarded using the set of local upper bounds $\mathcal{L_{LUB}}$, which is fixed after applying Algorithm 1 and contains more points then. Hence, the discarding test is more precise now and might allow us to discard additional boxes.

In the following algorithm we describe this procedure in detail.

Algorithm 5 Elimination step

INPUT: (MOMICP), $\mathcal{L_S}^1, \mathcal{L_{NS}}^1$, $\mathcal{L_{PNS}}$, $\mathcal{L_{LUB}}$
OUTPUT: $\mathcal{L_S}^2, \mathcal{L_{NS}}^2$

1: Set $\mathcal{L_S}^2 \leftarrow \varnothing$, $\mathcal{L_{NS}}^2 \leftarrow \mathcal{L_{NS}}^1$.
2: Set $\mathcal{F} \leftarrow$ true.
3: **for** $\tilde{B} \in \mathcal{L_S}^1$ **do**
4:     Apply Algorithm 3 to $\tilde{B}$.
5:     **if** $\mathcal{D}$ **then**
6:         Set $\mathcal{L_{NS}}^2 \leftarrow \mathcal{L_{NS}}^2 \cup \{\tilde{B}\}$.
7:     **else**
8:         Set $\mathcal{L_S}^2 \leftarrow \mathcal{L_S}^2 \cup \{\tilde{B}\}$.
9:     **end if**
10: **end for**

4.3 Decrease box width

In this section, we introduce a method that we can apply after Algorithm 5 in order to obtain a more precise approximation of the efficient set E of (MOMICP) than $\mathcal{L}_\mathcal{S}^2$. The idea is to split the boxes $\tilde{B} \in \mathcal{L}_\mathcal{S}^2$ till it holds $\text{wid}(\tilde{B}) \leqslant \delta$ for all boxes $\tilde{B}$ for a given $\delta \in \mathbb{R}$ with $0 < \delta < 1$. Therefore, we bisect the boxes by applying the bisection step and additionally check whether the respective subboxes can be discarded. This procedure is similar to Algorithm 1.

In the following algorithm we describe this procedure in detail.

Algorithm 6 Decrease box width

INPUT: (MOMICP), $\mathcal{L_S}^2$,$\mathcal{L_{NS}}^2$, $\mathcal{L_{PNS}}$, $\mathcal{L_{CUB}}$, $\delta \in \mathbb{R}$ with $0 < \delta < 1$
OUTPUT: $\mathcal{L_S}^3$,$\mathcal{L_{NS}}^3$

1: Set $\mathcal{L_S}^3 \leftarrow \varnothing$, $\mathcal{L_{NS}}^3 \leftarrow \mathcal{L_{NS}}^2$.
2: Set $\mathcal{F} \leftarrow$ true.
3: Set $\mathcal{L_W} \leftarrow \mathcal{L_S}^2$.
4: **while** $\mathcal{L_W} \neq \varnothing$ **do**
5:     Select a box $\tilde{B} \in \mathcal{L_W}$ via the selection rule.
6:     Set $\mathcal{L_W} \leftarrow \mathcal{L_W} \backslash \{\tilde{B}\}$.
7:     Bisect $\tilde{B}$ by applying the bisection step into $\tilde{B}^1$ and $\tilde{B}^2$.
8:     **for** $j = 1, 2$ **do**
9:        Apply Algorithm 3 to $\tilde{B}^j$.
10:        **if** $\mathcal{D}^j \vee \tilde{B}^j$ does not fulfill the necessary feasibility condition **then**
11:           Set $\mathcal{L_{NS}}^3 \leftarrow \mathcal{L_{NS}}^3 \cup \{\tilde{B}^j\}$.
12:        **else**
13:           **if** $\text{wid}(\tilde{B}^j) \leqslant \delta$ **then**
14:              Set $\mathcal{L_S}^3 \leftarrow \mathcal{L_S}^3 \cup \{\tilde{B}^j\}$.
15:           **else**
16:              Set $\mathcal{L_W} \leftarrow \mathcal{L_W} \cup \{\tilde{B}^j\}$.
17:           **end if**
18:        **end if**
19:     **end for**
20: **end while**

4.4 Enhanced algorithm and theoretical results

In this section, we introduce an enhanced algorithm that uses the ideas of Algorithm 4, Algorithm 5, Algorithm 6 in addition to the basic Branch-and-Bound algorithm for (MOMICP), referred to as Algorithm 1. Moreover, we prove some theoretical results regarding the enhanced algorithm.

Algorithm 7 Enhanced algorithm
INPUT: (MOMICP), imgp $\in \mathbb{N}$, $\varepsilon > 0$, $\delta \in \mathbb{R}$ with $0 < \delta < 1$
OUTPUT: $\mathcal{L}_{\mathcal{S}}{}^i$, $\mathcal{L}_{\mathcal{NS}}{}^i$ for $i \in \{1, 2, 3\}$, $\mathcal{L}_{\mathcal{PNS}}$, $\mathcal{L}_{\mathcal{LUB}}$
  1: Apply Algorithm 4.
  2: Set $\mathcal{F} \leftarrow$ false.
  3: Apply Algorithm 1.
  4: Set $\mathcal{F} \leftarrow$ true.
  5: Apply Algorithm 5.
  6: Apply Algorithm 6.

Remark 4.3. *Note that by setting $\mathcal{F} \leftarrow$ true in line 4 of Algorithm 7 we do not update $\mathcal{L}_{\mathcal{PNS}}$ and $\mathcal{L}_{\mathcal{LUB}}$ during Algorithm 5 and Algorithm 6 anymore. The idea behind this is that we already obtained many points $b \in \mathcal{L}_{\mathcal{PNS}} \subseteq f(B^{g,\mathbb{Z}})$ and hence many local upper bounds after applying Algorithm 4 and Algorithm 1. We might save some computational time by not adding even more points to $\mathcal{L}_{\mathcal{PNS}}$. Moreover, the list $\mathcal{L}_{\mathcal{PNS}}$ generated so far already contains enough information to approximate the set of nondominated points of (MOMICP) sufficiently accurate. Furthermore, we consider every box $\tilde{B} \in \mathcal{L}_{\mathcal{S}}{}^1$ again in Algorithm 5 and Algorithm 6 and apply the discarding test to this box. Therefore, we have to examine the respective hyperplanes for $\tilde{B}$. However, we already obtained some hyperplanes for $\tilde{B}$ during the discarding test in Algorithm 1. In order to save some computational time, we can save these hyperplanes during Algorithm 1 and recall them later. In that way we do not have to examine these hyperplanes again and only have to solve the respective optimization problems for some new points that were added to $\mathcal{L}_{\mathcal{LUB}}$ after $\tilde{B}$ was added to $\mathcal{L}_{\mathcal{S}}{}^1$.*

The following theorem proves the exactness of Algorithm 7, i.e. all $\mathcal{L}_{\mathcal{S}}{}^i$ for $i \in \{1, 2, 3\}$ are covers of the efficient set of (MOMICP). In addition, we prove a result regarding the precision of these covers.

Theorem 4.4. *After applying Algorithm 7, the sets $\mathcal{L_S}^1, \mathcal{L_S}^2, \mathcal{L_S}^3$ are covers of the efficient set E of (MOMICP). Furthermore, $\mathcal{L_S}^3$ is more precise than $\mathcal{L_S}^2$ and $\mathcal{L_S}^2$ is more precise than $\mathcal{L_S}^1$. Moreover, it holds $\underline{y'} = \overline{y'}$ for every box $\tilde{B} = [\underline{x'}, \overline{x'}] \times [\underline{y'}, \overline{y'}] \in \mathcal{L_S}^3$, i.e. the integer variables of $\tilde{B}$ are fixed.*

Proof. Considering Algorithm 5, it is obvious that $\bigcup \mathcal{L_S}^2 \subseteq \bigcup \mathcal{L_S}^1$ holds. Moreover, it is easy to observe that $\bigcup \mathcal{L_S}^3 \subseteq \bigcup \mathcal{L_S}^2$ holds by considering Algorithm 6. Due to Theorem 3.18 it is $E \subseteq \bigcup \mathcal{L_S}^1$. Because of Theorem 3.17 it also holds $E \subseteq \bigcup \mathcal{L_S}^2$. Morcover, Corollary 3.3, Theorem 3.17 and Corollary 3.6 imply $E \subseteq \bigcup \mathcal{L_S}^3$. So, the sets $\mathcal{L_S}^1, \mathcal{L_S}^2, \mathcal{L_S}^3$ are covers of the efficient set E of (MOMICP) and $\mathcal{L_S}^3$ is more precise than $\mathcal{L_S}^2$ and $\mathcal{L_S}^2$ is more precise than $\mathcal{L_S}^1$.

Now, assume that there is a box $\tilde{B} = [\underline{x'}, \overline{x'}] \times [\underline{y'}, \overline{y'}] \in \mathcal{L_S}^3$ with $\underline{y'} \neq \overline{y'}$. This implies that there is a $i \in \{1, ..., n\}$ with $\underline{y'}_i < \overline{y'}_i$. Due to our bisection method this implies $\underline{y'}_i \leqslant \overline{y'}_i + 1$. This implies $\text{wid}(\tilde{B}) \geqslant \overline{y'}_i - \underline{y'}_i \geqslant 1 > \delta$. This is a contradiction, because every box $B' \in \mathcal{L_S}^3$ fulfills $\text{wid}(B') \leqslant \delta$. Hence, $\underline{y'} = \overline{y'}$ holds. $\qquad\square$

5 Test instances and numerical results

We have implemented Algorithm 7 in MATLAB for $p = 2$. All tests
have been run on an Intel(R) Core (TM) i7-6700K CPU @ 4.00GHz
with 32GB RAM (2x DDR4-2399/16GB) on the operating system
Microsoft Windows 10 Pro version 10.0.17134 in MATLAB R2018a.
In the following we introduce the test instances we designed.

© Springer Fachmedien Wiesbaden GmbH, part of Springer Nature 2020
S. Rocktäschel, *A Branch-and-Bound Algorithm for Multiobjective Mixed-integer Convex Optimization*, BestMasters, https://doi.org/10.1007/978-3-658-29149-5_5

name	X	Y	$g(x,y)$	$f(x,y)$
balls1	$\begin{bmatrix}-2\\-2\end{bmatrix}, \begin{bmatrix}2\\2\end{bmatrix}$	$[-2,2]$	$x_1^2 + x_2^2 - 1$	$\begin{pmatrix} x_1 + y \\ x_2 - y \end{pmatrix}$
balls2	$\begin{bmatrix}-2\\-2\end{bmatrix}, \begin{bmatrix}2\\2\end{bmatrix}$	$[-2,2]$	$x_1^2 + x_2^2 - 1$	$\begin{pmatrix} x_1 + y \\ x_2 + \exp(-y) \end{pmatrix}$
parabolas1	$[-2,2]$	$[-2,2]$	0	$\begin{pmatrix} x + y \\ x^2 - y \end{pmatrix}$
parabolas2	$[-2,2]$	$[-2,2]$	$(x+2)^2 + (y+2)^2 - 16$	$\begin{pmatrix} x + y \\ x^2 - y \end{pmatrix}$
parabolas3	$[-2,2]$	$[-2,2]$	$(x-2)^2 + (y-2)^2 - 16$	$\begin{pmatrix} x + y \\ x^2 - y \end{pmatrix}$
points1	$[-2,2]$	$[-2,2]$	0	$\begin{pmatrix} x^2 \\ x^2 + (y - \frac{1}{2})^2 \end{pmatrix}$
points2	$[-2,2]$	$[0,1]$	$x^2 + (y - \frac{1}{2})^2 - \frac{1}{4}$	$\begin{pmatrix} x \\ x^2 \end{pmatrix}$
triangles1	$\begin{bmatrix}-2\\-2\end{bmatrix}, \begin{bmatrix}2\\2\end{bmatrix}$	$[-2,2]$	$\begin{pmatrix} -x_2 + \frac{3}{4}x_1 - \frac{1}{2} \\ x_2 - \frac{4}{3}x_1 - \frac{2}{3} \end{pmatrix}$	$\begin{pmatrix} x_1 - y \\ x_2 + y \end{pmatrix}$
triangles2	$\begin{bmatrix}-2\\-2\end{bmatrix}, \begin{bmatrix}2\\2\end{bmatrix}$	$[-2,2]$	$\begin{pmatrix} -x_2 - 3x_1 - 4 \\ -x_2 - \frac{1}{3}x_1 - \frac{4}{3} \end{pmatrix}$	$\begin{pmatrix} x_1 - y \\ x_2 + y \end{pmatrix}$

triangles3	$\left[\begin{pmatrix}-2\\-2\end{pmatrix}, \begin{pmatrix}2\\2\end{pmatrix}\right]$	$[-2, 2]$	$\begin{pmatrix} -x_2 + \frac{3}{4}x_1 - \frac{1}{2} \\ x_2 - \frac{4}{3}x_1 - \frac{2}{3} \\ -x_2 - x_1 - 3 - y \end{pmatrix}$	$\begin{pmatrix} x_1 - y \\ x_2 + y \end{pmatrix}$
triangles4	$\left[\begin{pmatrix}-2\\-2\end{pmatrix}, \begin{pmatrix}2\\2\end{pmatrix}\right]$	$[-2, 2]$	$\begin{pmatrix} -x_2 + \frac{3}{4}x_1 - \frac{1}{2} \\ x_2 - \frac{4}{3}x_1 - \frac{2}{3} \\ -x_2 - x_1 - 3 - 2y \end{pmatrix}$	$\begin{pmatrix} x_1 - y \\ x_2 + y \end{pmatrix}$
triangles5	$\left[\begin{pmatrix}-2\\-2\end{pmatrix}, \begin{pmatrix}2\\2\end{pmatrix}\right]$	$[-2, 2]$	$\begin{pmatrix} -x_2 + \frac{3}{4}x_1 - \frac{1}{2} \\ x_2 - \frac{4}{3}x_1 - \frac{2}{3} \\ -x_2 - x_1 - 3 - 2y \end{pmatrix}$	$\begin{pmatrix} x_1 - 2y \\ x_2 + 2y \end{pmatrix}$
highdim	$\left[\begin{pmatrix}-1\\-1\end{pmatrix}, \begin{pmatrix}1\\1\end{pmatrix}\right]$	$[-2e^8, 2e^8]$	$\begin{pmatrix} -\sum_{i=1}^{8} y_i \\ x_1 - x_2 \\ y_1 - y_4 \\ y_5 - y_2 \\ y_3 - y_6 \\ y_7 - y_8 \end{pmatrix}$	$\begin{pmatrix} x_1^2 - x_2 + \sum_{i=1}^{3} y_i \\ x_2^2 - x_1 + \sum_{i=4}^{6} y_i \end{pmatrix}$

Table 5.1: test instances

Furthermore, we have set $\varepsilon = \delta = 0.1$ for all instances and we have used different choices for imgp for which we compare the performance of the algorithm later in this section. Also note that all instances fulfill Assumption 2.9. Despite that, certain properties do not hold for some instances due to the mixed-integer constraints. For example the feasible sets $B^{g,\mathbb{Z}}$ are nonconvex and even not connected. Furthermore, the image sets $f(B^{g,\mathbb{Z}})$ are not necessarily connected and $f(B^{g,\mathbb{Z}}) + \mathbb{R}_+^p$ is not necessarily convex. These are properties that are often assumed to hold in multiobjective optimization.

The idea was to design test instances with efficient sets and nondominated sets that are easy to determine analytically but still illustrate these 'difficulties' that arise in multiobjective mixed-integer optimization. The underlying idea of the test instance 'highdim' was to test the algorithm on an instance with a high dimension of the pre-image space $r = 10$.

Plots that illustrate the points in $\mathcal{L}_{\mathcal{PNS}}$ after the termination of the algorithm with imgp $= 0$ colored black and image points in $f(B^{g,\mathbb{Z}})$ colored grey can be found in Appendix 7. We can see that we obtain a good approximation of the nondominated set using Algorithm 7. Note that it is not easy to illustrate the obtained covers $\mathcal{L}_S^{3}$ and the efficient sets E due to the integer constraints and the fact that $r > 2$ holds for many of our test instances. Therefore, we will not compare them here. In the following we compare how different choices of imgp for the preinitialization affect the performance of the algorithm. Therefore, we consider iteration counts and computational time needed. Table 5.2 lists the iteration counts. Hereby N denotes the number of points in $f(B^{g,\mathbb{Z}})$ we obtained during the preinitialization, J denotes the iteration count of Algorithm 1, K denotes the iteration count of Algorithm 5 and L denotes the iteration count of Algorithm 6.

name	imgp	N	J	K	L
	0	0	319	280	810
balls1	3000	120	301	246	811
	5000	195	276	222	842
	0	0	257	215	675
balls2	3000	120	239	190	690
	5000	195	239	188	691
	0	0	85	78	194
parabolas1	3000	3005	8	9	281
	5000	5005	8	9	279
	0	0	74	65	157
parabolas2	3000	2100	10	8	228
	5000	3500	10	8	228
	0	0	45	41	150
parabolas3	3000	2100	6	7	193
	5000	3500	6	7	191
	0	0	1	2	100
points1	3000	3005	1	2	100
	5000	5005	1	2	100
	0	0	3	3	22
points2	3000	2	3	3	22
	5000	2	3	3	22
	0	0	12	5	41
triangles1	3000	150	12	5	41
	5000	280	12	5	41
	0	0	134	75	1351
triangles2	3000	410	121	73	1351
	5000	785	123	74	1352

	0	0	26	12	233
triangles3	3000	138	24	13	218
	5000	262	24	14	219
	0	0	27	6	148
triangles4	3000	126	26	5	149
	5000	243	26	6	150
	0	0	43	16	283
triangles5	3000	126	49	19	276
	5000	243	46	18	281

Table 5.2: iteration counts

We observe that as expected, we need less iterations J and K for most test instances when using the preinitialization. However, we then still have to decrease the box width in Algorithm 6 and therefore need more iterations for L in most cases. When investigating the values for 'parabolas1', 'parabolas2' and 'parabolas3' we can observe this relation very good. Summarizing it seems that we tend to have less overall iterations when using the preinitialization as expected. However, we might need more computational time for each iteration when using the preinitialization. Hence, we consider the computational times in the following table. Hereby, t_{init} denotes the time needed for Algorithm 4, t_1 denotes the time needed for Algorithm 1, t_2 denotes the time needed for Algorithm 5, t_3 denotes the time needed for Algorithm 6, t_{visual} denotes the time needed for plots and t_{total} denotes the total time needed for executing Algorithm 7.

name	imgp	N	t_{init}/s	t_1/s	t_2/s	t_3/s	t_{visual}/s	t_{total}/s
balls1	0	0	0.05	13.56	1.81	7.82	0.52	23.77
	3000	120	0.06	11.60	1.63	8.12	0.29	21.71
	5000	195	0.04	10.39	1.54	8.53	0.27	20.78
balls2	0	0	0.06	9.83	1.32	6.37	0.31	17.89
	3000	120	0.05	9.27	1.30	6.83	0.30	17.75
	5000	195	0.05	8.99	1.24	6.71	0.28	17.26
parabolas1	0	0	0.04	3.15	2.38	1.74	0.98	8.29
	3000	3005	1.31	21.60	18.31	58.65	1.00	100.88
	5000	5005	3.62	57.70	41.43	122.55	1.22	226.52
parabolas2	0	0	0.04	2.97	1.88	1.67	1.06	7.62
	3000	2100	0.96	17.21	14.85	49.28	0.91	83.21
	5000	3500	2.45	45.55	33.02	100.56	0.92	182.51
parabolas3	0	0	0.03	2.18	1.31	1.80	0.87	6.19
	3000	2100	0.54	7.17	6.97	25.64	0.83	41.16
	5000	3500	1.35	16.02	14.47	48.64	0.84	81.32
points1	0	0	0.03	0.28	0.29	0.81	0.68	2.10
	3000	3005	0.17	0.32	0.25	0.76	0.68	2.18
	5000	5005	0.23	0.28	0.26	0.81	0.68	2.26
points2	0	0	0.04	0.36	0.26	0.39	0.54	1.58
	3000	2	0.04	0.36	0.36	0.37	0.55	1.68
	5000	2	0.03	0.36	0.26	0.37	0.55	1.56

triangles1	0	0	0.05	0.67	0.02	0.48	0.30	1.53
	3000	150	0.05	0.61	0.02	0.59	0.29	1.55
	5000	280	0.03	0.62	0.02	0.64	0.29	1.60
triangles2	0	0	0.04	6.54	1.40	24.81	0.27	33.06
	3000	410	0.05	6.09	1.47	24.88	0.28	32.78
	5000	785	0.05	6.05	1.37	24.48	0.30	32.25
triangles3	0	0	0.04	1.56	0.14	4.36	0.28	6.38
	3000	138	0.05	1.42	0.13	4.17	0.28	6.05
	5000	262	0.04	1.41	0.12	4.04	0.34	5.94
triangles4	0	0	0.04	1.50	0.06	2.75	0.28	4.64
	3000	126	0.05	1.47	0.04	2.86	0.27	4.69
	5000	243	0.04	1.53	0.06	2.85	0.28	4.75
triangles5	0	0	0.04	2.48	0.17	5.27	0.27	8.24
	3000	126	0.05	2.79	0.20	5.07	0.29	8.39
	5000	243	0.04	2.66	0.20	5.31	0.27	8.48

Table 5.3: computational times

We observe that the preinitialization saves some computational time for some test instances like 'balls1' and 'balls2', but this is neglectable when considering the total times needed for 'parabolas1', 'parabolas2' and 'parabolas3'. We need extensively more computational time to solve these problems when using the preinitialization despite less iterations needed. This suggests that a preinitialization is not necessary for Algorithm 7 and even hindering, because we might need much more computational time to solve an optimization problem than without the preinitialization.

Note that we did not obtain results for the test instance 'highdim', because we stopped the test on this instance after 24 hours. So we can see that the algorithm solves these low dimensional test instances rather quickly and without a need of a preinitialization, but it definitely has its limits regarding higher dimensional pre-image spaces.

6 Outlook and further possible improvements

In this Chapter, we discuss an extension of the proposed algorithm to the nonconvex case. Therefore, we introduce the concept of convex underestimators. As we have seen in Example 2.13, the assumption of convexity of f and g for (MOMICP) in Assumption 2.9 can be very restricting. However, the concept of convex underestimators will allow us to drop the assumption of convexity of f. Hence, we are then able to consider a larger class of optimization problems. In the following, we outline how the concept of convex underestimators allows us to drop the assumption of convexity of f in Assumption 2.9. At first, we introduce the definition of convex underestimators.

Definition 6.1. *Let $B \in \mathbb{R}^r$ be a box for $r \in \mathbb{N}$ and $h \colon B \to \mathbb{R}$ a given function. A function $\underline{h} \colon B \to \mathbb{R}$ is called a convex underestimator of h (on B), if $\underline{h}$ is convex and $\underline{h}(b) \leqslant f(h)$ holds for all $b \in B$.*

Considering (MOMICP) with an objective function f that is not necessarily convex, we observe that the optimization problems $(\mathrm{HPOP}_z(\tilde{B}))$ for $z \in \mathcal{L}_{\mathcal{CUB}}$ that we solve during Algorithm 7 are not easy to solve anymore. We would need a global optimization approach in order to solve these problems due to f not being convex. However, solving such problems can take some time and since we have to solve many of these optimization problems this might result in a huge increase in computational time overall.

Instead, our idea is to determine a convex underestimator $\underline{f}_i$ of f_i for all $\in \{1, ..., p\}$ on the subbox $\tilde{B}$ of B we are considering for $(\mathrm{HPOP}_z(\tilde{B}))$. Since

© Springer Fachmedien Wiesbaden GmbH, part of Springer Nature 2020
S. Rocktäschel, *A Branch-and-Bound Algorithm for Multiobjective Mixed-integer Convex Optimization*, BestMasters, https://doi.org/10.1007/978-3-658-29149-5_6

f_i is twice continuously differentiable this can easily be done as shown in [14]. Then we replace the constraint $f(b) \leqslant z + t \cdot e$ of $(\text{HPOP}_z(\tilde{B}))$ with $\underline{f}(b) \leqslant z + t \cdot e$ which makes this problem easier solvable since $\underline{f}$ is convex. It can be shown that we still obtain lower bounds for $f(\tilde{B}^g)$ in that way [16]. These can then be used as lower bounds for $f(\tilde{B}^{g,\mathbb{Z}})$.

The drawback of this method is that the lower bounds are less tight. However, with decreasing box width $\text{wid}(\tilde{B})$ the bounds get tighter since $\underline{f}$ is a better approximation for f then, see [14].

Summarizing, we are able to drop the assumption of convexity of f in Assumption 2.9, but since we obtain bounds that are not very tight for nonconvex f, we assume that solving (MOMICP) would need more iterations and hence more computational time then. However, this would allow us to solve optimization problems like (P_y) for example, if we do not have to consider any constraints $(\tilde{g} \equiv 0)$.

Note that the theoretical results of this book also hold for $p > 2$. Hence, a further improvement could be to revisit the implementation of Algorithm 7 in order to generalize it to being able to solve (MOMICP) for all $p \geqslant 2$.

7 Conclusion

In this book, we have considered multiobjective mixed-integer convex optimization problems. We introduced basic definitions and concepts of multiobjective optimization. We derived a basic Branch-and-Bound algorithm for solving (MOMICP) by using approaches from global multiobjective (continuous) optimization [16]. We have described the main steps of this algorithm and proven exactness. We have enhanced this basic algorithm by introducing modifications that can save computational time or return a more precise cover of the efficient set E of (MOMICP). The final algorithm was implemented in MATLAB. We have tested the algorithm on several instances which have been designed by us. We have discussed the impact of the modifications on computational time and precision of the cover of E. Finally, we have outlined further steps that could be done in order to save more computational time, obtain a more precise cover of E and generalize our theoretical results and the final algorithm in order to handle optimization problems (MOMICP) with not necessary convex objective functions f.

© Springer Fachmedien Wiesbaden GmbH, part of Springer Nature 2020
S. Rocktäschel, *A Branch-and-Bound Algorithm for Multiobjective Mixed-integer Convex Optimization*, BestMasters, https://doi.org/10.1007/978-3-658-29149-5_7

Bibliography

[1] ABHISHEK, K. ; LEYFFER, S. ; LINDEROTH, J. : FilMINT: An outer approximation-based solver for convex mixed-integer nonlinear programs. In: *INFORMS Journal on computing* 22 (2010), Nr. 4, S. 555–567

[2] BELOTTI, P. ; KIRCHES, C. ; LEYFFER, S. ; LINDEROTH, J. ; LUEDTKE, J. ; MAHAJAN, A. : Mixed-integer nonlinear optimization. In: *Acta Numerica* 22 (2013), S. 1–131

[3] BENSON, H. P.: An outer approximation algorithm for generating all efficient extreme points in the outcome set of a multiple objective linear programming problem. In: *Journal of Global Optimization* 13 (1998), Nr. 1, S. 1–24

[4] BONAMI, P. ; CORNUÉJOLS, G. ; LODI, A. ; MARGOT, F. : A feasibility pump for mixed integer nonlinear programs. In: *Mathematical Programming* 119 (2009), Nr. 2, S. 331–352

[5] CACCHIANI, V. ; DÂĂŽAMBROSIO, C. : A branch-and-bound based heuristic algorithm for convex multi-objective MINLPs. In: *European Journal of Operational Research* 260 (2017), Nr. 3, S. 920–933

[6] EHRGOTT, M. : *Multicriteria Optimization.* Springer, 2005

[7] EICHFELDER, G. : *Adaptive Scalarization Methods in Multiobjective Optimization.* Springer, 2008

[8] FERNÁNDEZ, J. ; TÓTH, B. : Obtaining the efficient set of nonlinear biobjective optimization problems via interval branch-and-bound methods. In: *Computational Optimization and Applications* 42 (2009), Nr. 3, S. 393–419

© Springer Fachmedien Wiesbaden GmbH, part of Springer Nature 2020
S. Rocktäschel, *A Branch-and-Bound Algorithm for Multiobjective Mixed-integer Convex Optimization*, BestMasters, https://doi.org/10.1007/978-3-658-29149-5

[9] GÜNLÜK, O. ; LEE, J. ; WEISMANTEL, R. : MINLP strengthening for separable convex quadratic transportation-cost UFL. In: *IBM Res. Report* (2007), S. 1–16

[10] GUROBI OPTIMIZATION, L. : *Gurobi Optimizer Reference Manual.* http://www.gurobi.com. Version: 2018

[11] JAHN, J. : *Vector optimization.* Springer, 2009

[12] KLAMROTH, K. ; LACOUR, R. ; VANDERPOOTEN, D. : On the representation of the search region in multi-objective optimization. In: *European Journal of Operational Research* 245 (2015), Nr. 3, S. 767–778

[13] LEE, J. ; LEYFFER, S. : *Mixed integer nonlinear programming.* Bd. 154. Springer Science & Business Media, 2011

[14] MARANAS, C. D. ; FLOUDAS, C. A.: Global minimum potential energy conformations of small molecules. In: *Journal of Global Optimization* 4 (1994), Nr. 2, S. 135–170

[15] NEUMAIER, A. : *Interval methods for systems of equations.* Bd. 37. Cambridge university press, 1990

[16] NIEBLING, J. ; EICHFELDER, G. : A Branch-and-Bound based Algorithm for Nonconvex Multiobjective Optimization. In: *Preprint-Series of the Institute for Mathematics* (2018)

Plots referring to the numerical tests

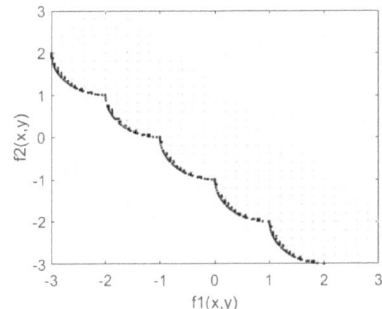 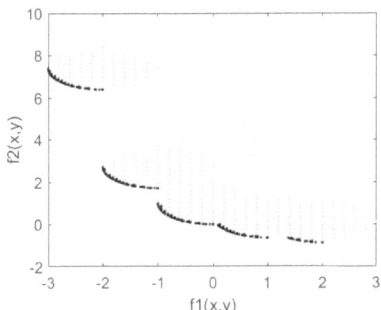

Figure 1: image set and $\mathcal{L}_{\mathcal{PNS}}$ of balls1

Figure 2: image set and $\mathcal{L}_{\mathcal{PNS}}$ of balls2

© Springer Fachmedien Wiesbaden GmbH, part of Springer Nature 2020
S. Rocktäschel, *A Branch-and-Bound Algorithm for Multiobjective Mixed-integer Convex Optimization*, BestMasters, https://doi.org/10.1007/978-3-658-29149-5

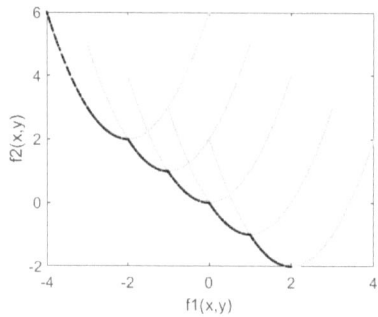

Figure 3: image set and $\mathcal{L}_{\mathcal{PNS}}$ of parabolas1

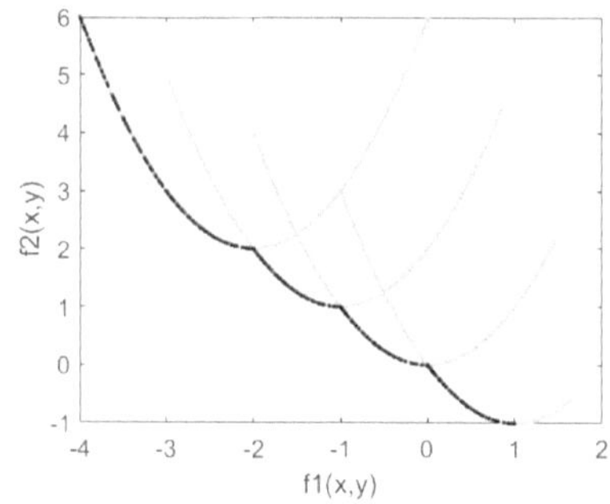

Figure 4: image set and $\mathcal{L}_{\mathcal{PNS}}$ of parabolas2

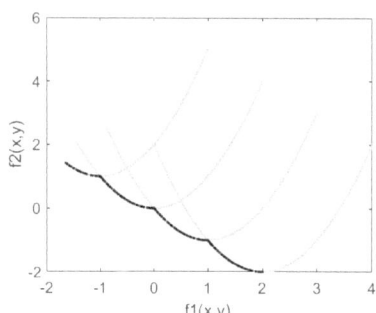
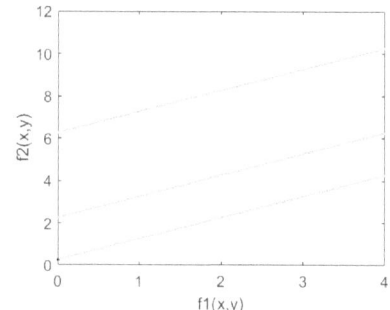

Figure 5: image set and $\mathcal{L}_{\mathcal{PNS}}$ of parabolas3

Figure 6: image set and $\mathcal{L}_{\mathcal{PNS}}$ of points1

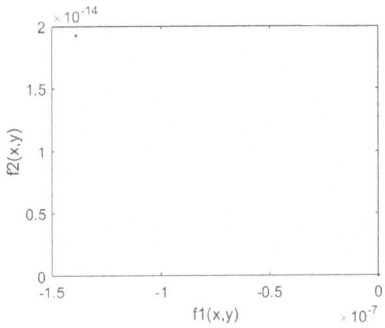

Figure 7: image set and $\mathcal{L}_{\mathcal{PNS}}$ of points2

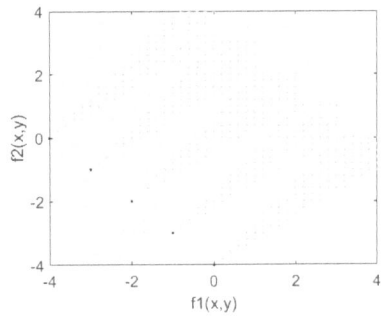

Figure 8: image set and $\mathcal{L}_{\mathcal{PNS}}$ of triangles1

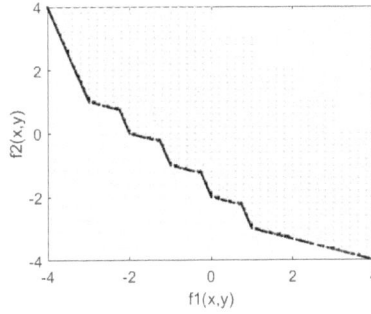

Figure 9: image set and $\mathcal{L}_{\mathcal{PNS}}$ of triangles2

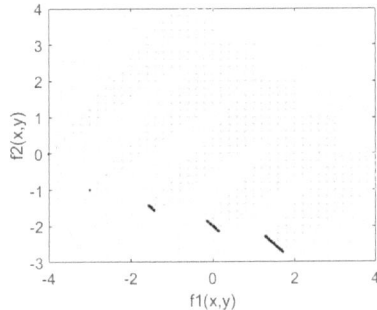

Figure 10: image set and $\mathcal{L}_{\mathcal{PNS}}$ of triangles3

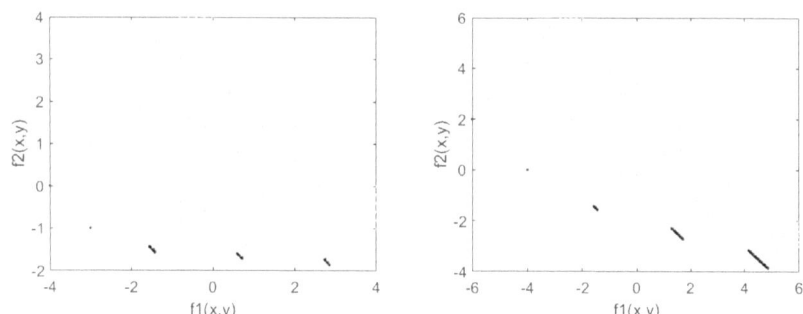

Figure 11: image set and $\mathcal{L}_{\mathcal{PNS}}$ of triangles4

Figure 12: image set and $\mathcal{L}_{\mathcal{PNS}}$ of triangles5